AMPUTATION ON REQUEST
Alex Mensaert

Why people want to become disabled
& Why people are attracted to disability

Copyright: Alex Mensaert-Hedera – 2011
www.alexmensaert.com
ISBN: 978-1-257-92926-9

By way of introduction.

Very difficult it would be if we still would live in the *silent years*, were it was one big taboo to talk about *apotemnophilia* and *acrotomophilia*, or to say it in different words; People who like amputees and those who are attracted to them, or people who want to become an amputee.

First of all, who I am; A guy from Belgium, a little country in Europe, and most of all important, an amputee. No legs, double above the knee amputated and one arm, left below the elbow amputated.

I became an amputee in 1997, in an accident. After that I was bored so much, from not knowing what to do the whole days as an amputee that I started with making websites. The internet that time wasn't so big as it is now, but soon I came in contact with people that told me that they found me very attractive because I had those stumps. Indeed, those people found me more beautiful with my amputations then they had found me if I had been a non-amputee.

More and more I came in contact with different people who were attracted to me because my amputations were so attractive for them.

My interest in the subject became bigger day by day, and I liked the idea that a lot of people on the world were interested in me, just because my leg was replaced by a stump. That time I was a left below the knee amputee only.

I decided to make a website, named Ampulove. Based on my story and for people interested in amputees. And surprised I was for sure, when I found out that so many visitors found the way to Ampulove. That time there were about two thousand visitors each day on the site.

Later some people asked my address and started to send me different amputee pictures from all over the world, with the request to put them on the internet.

After placing an amount of amputee pictures on internet, a different group of people started to contact me. People that told me that they were jealous on me and the people on the pictures I had placed online; That they wanted to be an amputee. I was surprised, but also very interested in their stories. I remember different stories from different people who contacted me during all the years and will try to tell, a few of them, in this book.

After my accident I had a lot of problems with my other leg and with my right arm. Doctors decided later, during the next years to do more amputations. But with every amputation that was necessary to do, I never felt me really down or depressive. Always I wanted to proof the whole world that I could do things without legs and with only one arm. I wanted to show this blue planet that amputees without legs and one arm still can do everything, if not, more then some four-limbed people.

And I did. During the last fourteen years, after my first amputation I did so many things. I was a tattoo artlst, I had different stores, I made websites, and that time the Ampulove site was one of the biggest sites on the world when it came to the amputee subject.

The amputee subject…. I discovered that people who like amputees are named Acrotomophiles. Coming from the Greek. Near that the group that contacted me later, and were jealous on me and my stumps; Those people are called Apotemnophiles, also coming from Greek and means ´love for cutting´. I started to read articles on the internet about the two subjects and found out that Dr. J Money from New York was the first psychologist who did investigations in 1977 on apotemnophiles, also wannabes named, coming from: want to be. In this case, want to be an amputee.

Dr. Money said that about one on thirty thousand people are a wannabe. That is then the same almost as the group transsexuals housed on our blue planet. Also in their *world of strangeness* they are with about the same amount of people as apotemnophiles. About the group of acrotomophiles, people admired to stumps, I found out that about one on five hundred persons would be interested in people with an amputation. Acrotomophiles are also named as Devotees. Devoted to…. Stumps, amputation and even more. Not only devotees for amputees, but also devoted to more medical *fetishes*. Some of them are interested in wheelchair users, plaster casts, braces, blind people, and so much more different disabilities.

One person on five hundred, males and females, young and old are interested in stumps on the first place -interested in disabilities; A lot if you think about.

Now, looking at myself. After all those years of being a triple amputee; Maybe I need to consider myself as a kind of wannabe. A wannabe that became an apotemnophile after becoming an amputee. Every time a surgeon took of more from me, every time I really didn't mind he did, I was only interested in showing the world that I still would be able to do everything. Maybe some things in a different way, but I still would do everything like all the people I know, the people with two legs and two arms. Considering myself a wannabe or just some amputee who became very open minded for the group of human beings interested in stumps and disability.

My interest in the group of devotees and wannabes was on a high level after I became a legless and one armed man, and has grown so much, that I can say now after all those years, that I have more devotee and wannabe friends on the world then people who are not disabled.

I met during all the years a lot of amputees. Ex-wannabes and also amputees that never had heard from people who want to become an amputee by wish. I met different apotemno -and acrotomophiles from all over the world.

I went to the United Kingdom to inform psychiatrics about all I found about the deeply wish of some people who wanted one or more stumps. I was in many worldwide magazines with articles about me and how open minded I got for the, that time, *secret dark world* of *stump fetishism*.

After the conversations I had in my amputated life, with the group of people, were I'm writing a book about, I decided to publish all my knowledge, investigations and stories.

Years ago, it would had been very difficult to talk about this, to write about this subject. The first one who ever wrote a book about the apotemnophilia subject was Gregg M. Furth, PhD and an American psychologist and graduate of the C.G. Jung Institute in Zürich, who also happens to be an apotemnophile. Gregg, who I have known personally, died a few years ago.

The stories I mention in this book are all real stories, no fiction. Only that I changed the real names from some people, this because I don't want those people ending up in problems with their families or friends.

For me, as a person who is open minded to a lot of things, I only can say that it is sad to see how many people need to suffer, just because no one understand their feelings or special attractions. Also for that reason this open minded book about something that really exist but that some persons ignore.

A lot of wannabes that contacted me in the past, died from trying to become an amputee, but ended up in accidents where in they didn't wanted to lose life, but they wanted to lose a limb to restart a life, in the way they wanted... as an amputee. Unfortunately the biggest group of surgeons are not yet ready for this world of wannabes.

Alex Mensaert – 2011

* * *

Wannabe and Apotemnophilia

After the name of wannabes and want to be, apotemnophilia and Dr J. Money that was behind of all the first investigations on them, the conclusion was that being a wannabe was something sexual. Sexual? Indeed; A big group of apotemnophiles that contacted me during all the years confirmed me that every time they thought about being an amputee, they became *horny*. Not only my investigations mentioned a lot of wannabes that became hot with the ideas around being an amputee, but even Dr J. Money mentioned that in one of his early investigations.

Becoming an amputee by wish. Very strange if you think about it on the first place, but how deeper the investigation goes on, how more results are coming up.

One of the next wannabe investigations was with the English documentary *Horizon*. The investigator for the documentary contacted me and asked or I could help him with finding successful wannabes and people who wanted to become an amputee. He explained me that a surgeon, Dr. Smith in Scotland and the Falkirk hospital did legal amputations on wannabes after seeing a psychiatrist and psychologist, who both needed to agree with having the amputation done.

I went to the United Kingdom and met the psychiatrist and psychologist. They told me that finally the name of wannabe and apotemnophilia would be mentioned in the next medical book. What medical book they were talking about, I had no any idea.

After the *Horizon* documentary came out, the name of BID came up. BID means *body image disorder*. Also BDD, what means *body dysmorphic disorder*, a severe mental illness characterized by debilitating misperceptions that one appears disfigured and ugly.

The symptoms of BDD are; Obsessive thoughts about perceived appearance defects; Major depressive disorder symptoms; Suicidal ideation and more symptoms that have nothing to do with someone who want to become an amputee. After more worldwide investigations, producer Melody Gilbert, an award-winning and independent filmmaker, educator and a journalist from the St. Paul at Minnesota came out with the documentary – film ´*Whole*´ that premiered at the Los Angeles Film Festival in June 2003, and was broadcast on the

Sundance Channel and numerous TV stations around the world. Whole, a documentary about wannabes and different medical professionals that need to deal daily with wannabes from all over the world.

The name wannabe and apotemnophile transferred into BIID, what stands for - *Body integrity identity disorder*, and is a neurological and psychological disorder that makes sufferers feel that they would be happier living as an amputee. It is typically accompanied by the desire to amputate one or more healthy limbs to achieve that end.

A person with BIID wants one or more limbs amputated. While the official definition of BIID includes only a desire for amputation, Dr. Michael B. First, an author of the upcoming DSM-V (*Diagnostic and Statistical Manual of Mental Disorders*), who first defined BIID, has agreed in principle that BIID could include a need for other impairments, such as *paraplegia*, what is an impairment in motor or sensory function of the lower extremities, or partial paralysis of a limb.

A sexual motivation for being or looking like an amputee is called *apotemnophilia*. Most people with BIID don't report a sexual motivation, says the original BIID terms; What is so strange while so many wannabes told me in all the investigations and conversations I had with them, that they had sexual feelings with the idea of being an amputee. Not only the want to be, but also a big part of the successful apotemnophiles, who became an amputee by wish confirmed that their stump(s) are sexual involved and turns them on.

In addition, apotemnophilia should not be mistaken for acrotomophilia, which describes a person who is sexually attracted to other people who are already missing limbs. However, there does seem to be some relationship between the disorders, with some individuals exhibiting both conditions.

Today, no surgeons will treat BIID patients officially, by performing the desired amputations. Some act out their desires, pretending they are amputees using prostheses and other tools to ease their desire to be one. Some sufferers have reported to the media or by interview over the telephone with researchers that they have resorted to self-amputation of a "superfluous" limb, for example by allowing a train to run over it, or by damaging the limb so badly that surgeons will have to amputate it. However, the medical literature records few, if any, cases of actual self amputation. Often the obsession is with one specific limb. A patient might say, for example, that they "do not feel complete" while they still have a left leg. However, BIID does not simply involve amputation. It involves any wish to significantly alter body integrity.

Some people suffer from the desire to become paralyzed, blind, deaf, use orthopedic appliances such as leg-braces, etc. Some people spend time pretending they are an amputee by using crutches and wheelchairs at home or in public; in the BIID community, this is called a "pretender."

Exact causes for BIID are unknown. One theory states that the psyche of a child seeing an amputee, may imprint on this body image as an "ideal." Another popular theory suggests that a child who feels unloved may believe that becoming an amputee will attract sympathy and love.

The biological theory is that BIID is a neuro-psychological condition in which there is an anomaly in the cerebral cortex relating to the limbs; cf. *Proprioception*. (coming from Latin *proprius*, meaning "one's own" and perception, is the sense of the relative position of neighboring parts of the body.) If the condition is neurological, it could be conceptualized as a congenital form of *somatoparaphrenia* (Somatoparaphrenia is a type of monothematic delusion where one denies ownership of a limb or an entire side of one's body. As an example, a patient would believe that her or his own arm would belong to the doctor, or that another patient left it behind), a condition that often follows a stroke that affects the parietal lobe.

Since the right side of the inferior-parietal lobule—which is directly related with proprioception—is significantly smaller in men than women, a malfunction of this area could potentially explain not only why men are much more likely to have BIID, but also why requests for amputations most often concern left-side limbs. (The right side of the brain controls the left side of the body and vice versa.) If the condition is similar to somatophrenia, it could have the same "cure"—vestibular caloric stimulation. In simple terms it involves squirting cold water in the patient's right ear.

Symptoms of BIID sufferers are often keenly felt. The sufferer feels incomplete with four limbs, but his confident with amputation will fix this. The sufferer knows exactly what part of which limb should be amputated to relieve the suffering. The sufferer has intense feelings of envy toward amputees. They often pretend, both in private and in public, that they are an amputee. The sufferer recognizes the above symptoms as being strange and unnatural. They feel alone in having these thoughts, and don't believe anyone could ever understand their urges. They may try to injure themselves to require the amputation of that limb. They generally are ashamed of their thoughts and try to hide them from others, including therapists and health care professionals.

Near the official explanation of BIID I need to disagree with so many things. When we go deeper into the point that the majority of BIID sufferers are official white middle-aged males, and that the most common request is an above-the-knee amputation of the left leg.

During all the investigations and conversations I did with so many wannabes on our planet I realized that

95% of the wannabes had for the first time their wannabe feeling on a very young age. The most of them when they were about six till ten years young. Mostly after seeing an amputee in the street, a movie or somewhere else. The first contact with the fact that there are amputees on the world was also basically the first contact with the impulse to become an amputee; The maybe first jealous feeling like ´Why I am not that way, I want to be like him or her´.

If the original texts mention about the white middle aged man, then this is a sad thing, knowing that there are so many wannabes who are black, dark and even females.

Years ago I was in contact with a female wannabe from the United States. She absolutely wanted to become a double above the knee amputee. She went through the investigations of the psychiatrist and psychologist in the United Kingdom and the Horizon documentary, but they didn't agreed with the fact that this woman was a real wannabe. Later, the same woman who was once in front of those two medical people became an amputee. First left above the knee, afterwards right above the knee, what made her a double above the knee amputee. A dream she had already since childhood.

Another female wannabe from the United Kingdom, I was in contact with years ago, told me trough the internet that time, that she absolutely wanted her toes off from one foot. She was very young and worked in an hospital as a nurse. One night when I checked my email she wrote me that she had cut one of her own toes off using ice cubes to not feel the pain. Or it was the truth what she wrote me that time, I only discovered a few days later, with an other email that included some pictures from her and the self-amputation she had done. She went to the hospital and told the doctors that a heavy butchers knife, she had at home fell on her foot while she was cooking with naked feet in the kitchen. Just a few stories of female amputees. Amputees because they were wannabes.

An other story is from S.H. a female amputee who was a wannabe, and wrote me a few years ago a very interesting text. A story that makes us all clear that there are indeed female wannabes:

—In the two years since I first gained access to the Internet and World Wide Web, I have seen a fair amount of material concerning so-called "wannabes," individuals who harbor an intense desire to undergo amputation of a limb or limbs. Some of this material, including fictional stories and web sites that use digitally altered photographs that purport to show gorgeous women who have attained voluntary amputation, are so disconnected from reality as to verge on the hilarious. Some more-serious items are simply misinformed, either based on speculation or on psychological evaluation of wannabes who suffered from severe mental disturbances that may or may not have been connected to their desire for limb removal. None of it has seemed particularly relevant to the wannabe phenomenon as I understand it, and I think I understand it pretty well, at least so far as it pertains to me.

I am one-legged, female, married and 28 years old at this writing. I underwent voluntary left upper-thigh amputation when I was 16. This occurred with the consent of my parents and on the recommendation of my psychiatrist, following almost two years of therapy and evaluation, as treatment for an obsession with becoming an amputee that was interfering with every other aspect of my life. Or in other words, I was a wannabe, and was allowed to have an amputation because my doctors correctly believed that would be less disabling than my desire to have it. I consider myself extremely fortunate in obtaining the aid of physicians who saw my condition as a valid reason for surgery, as I was in having parents open-minded and loving enough to accept such a strange need on the part of their daughter and permit her to satisfy it.

I offer no explanation for the source of my desire for amputation. From the beginning of my school career I excelled academically and athletically. I made friends easily and was popular with my classmates. I was not sexually aroused by seeing or thinking about amputees, or by fantasies about being an amputee myself. I had no fetish for prosthetics or crutches, having never seen the former and feeling no desire to use the latter in the absence of an amputation. I was repulsed by thoughts of pain or being inferior to other people. I do not believe I had conscious or unconscious wishes to lower the standards of performance by which I was judged, attract the attention or pity of my peers, or make myself sexier, nor do I see myself as masochistic. Moreover, all such explanations were rejected by my psychiatrist prior to my surgery. This is not to discount the possibility that other wannabes might have those motivations. It would be understandable if some did. My point here is that any all-inclusive theory, if there is one, cannot be valid unless it accounts for people like myself who were not so motivated.

- I cannot totally reject the idea that exhibitionism was involved in my desire for amputation, because I am an exhibitionist by nature. In addition to being a typical child "show off" who enjoyed demonstrating my academic and athletic skills, from the time of puberty I found myself tremendously excited by the prospect or experience of making my female attributes available for masculine viewing. Beginning at age twelve or thirteen, I regularly exercised in the nude in front of my bedroom window while a male neighbor watched me from his house, and on numerous occasions I disrobed for various neighborhood boys, allowing them to see my breasts and vulva, all the while enjoying my body's power to give them erections. I found doing these things to be extremely arousing, and after each such episode I would retreat to my bathtub and masturbate to orgasm. Thus I suppose it might be that a belief I would be similarly aroused by being seen as an amputee (amputation being even more taboo than nudity) played a part in my wanting to be one. - On the other hand, I do not recollect having such thoughts on a conscious level, and I have never experienced sexual excitement at being observed by males while I was on one leg and crutches in the twelve years I've been capable of appearing publicly in that fashion. In this context I think it is significant that I frequently go without panties because that gives me a certain excitement in public even though I know my lack of underwear is not visible through my clothing, but am not aroused at all by having men look at me when I am wearing a swimsuit that leaves my stump completely exposed while concealing my genitalia. My exhibitionism does not seem to operate at an overtly sexual level where my amputation is concerned, but was and remains centered on the normal parts of my female anatomy. - There still remains the possibility I thought being an amputee would allow me to show off in a way that was extreme yet socially acceptable by simply going about my business on crutches, presenting other people with an exotic appearance rather than an act of skill, thereby obtaining the same satisfaction I got from doing well at sports and schoolwork. (As it turns out, such a belief might have been justified, because I do in fact obtain non-sexual gratification from being seen as able to perform everyday tasks as well as or better than most able-bodied people in spite of my obvious physical limitation, and I accept as fact that others would not find my performances so impressive were I not visibly one-legged.) I do not remember thinking this, but I mention it because it would be consistent with my enjoyment of showing off and it might have been something I accepted as a given without conscious thought. - As best I have been able to understand the feelings driving me to seek surgery, they seem to have fallen in three main areas: First, there is what I would categorize as a very strong but simple esthetic preference for the amputee form. Specifically, to my eye and all else being equal, a man or woman with one leg has always seemed more beautiful than a two-legged individual. I cannot explain the origin of this preference except to suggest it might have something to do with the asymmetry created by a single leg amputation. I do know I coveted this appearance for myself, in the heartfelt belief I would be more beautiful if I had it. - The second part of my desire comprised a raging curiosity about the sensations arising from amputation. For lack of a stronger word, I lusted to know what it felt like to have a stump instead of a leg. I wished to know all aspects of it with an intensity and urgency I cannot convey to anyone who has not experienced these same desires. - Finally, I wanted the constant added challenge of life on one leg, of being handicapped in the sporting sense of the term. I believed with unquestioning faith that my life would be fuller if the activities of daily life were more difficult to accomplish. But it is important to realize I was not interested in punishing myself or holding myself back. What excited me was not the added difficulty, but the prospect of overcoming it. I was not seeking an excuse for failure or mediocre performance. I wanted to do everything well. I merely wished to increase the satisfaction I would obtain from that. - One reason my psychiatrist felt justified in recommending surgery as suitable therapy for me was his understanding that these different aspects of my desire were not irrational in the context of the desired goal. Everything I wanted was attainable by and consistent with single leg amputation. I would have the altered form I thought most beautiful. I would experience the sensations normal to someone with a thigh stump. Most important, there was no reason to believe I could not perform at a superior level as an amputee. - Some who hold forth on the supposed drawbacks of satisfying a desire to lose a limb seem untrammeled by actual familiarity with real amputees. They begin with the assumption that amputation must result in reduced employability and earning potential, and go downhill from there.

The plain truth, however, is that the absence of one leg need not cause serious loss of function. Although it is not as trivial a disability as, say, 20/40 myopia, on the overall scale of things it rates as no more than a relatively minor inconvenience for a person in reasonably good health and physical condition living in the United States or other urbanized countries with highly-developed infrastructure and laws protecting the rights of the handicapped. The many single-leg amputees I have met who state they can still do anything they like are not deluding themselves or attempting to make light of a bad situation. They are presenting a realistic appraisal of their capabilities. They lead normal lives. They earn incomes sufficient to support themselves and their families. They do not become outcasts rejected by their communities. - In fact, a surprising number of amputees actually come to like their altered body images because of the advantages they confer in many social situations. A pinned-up empty trouser leg is an effective tool for obtaining special benefits and consideration, and amounts almost to a guarantee that one will everywhere be treated with politeness and respect, if not kindness. We should hardly be surprised that some with involuntarily acquired amputations come to enjoy the effect of their exotic appearance on the people they encounter. If this is true of amputees who had limb loss thrust upon them, how much more likely is it that a realized wannabe will enjoy and continue to enjoy his altered body? - It would be foolish to base our judgments of wannabes' expectations on social and economic conditions that no longer exist. This is not the 19th Century. Most amputees operate in the mainstream of society, not as second-class citizens relying on the sufferance and largesse of the able-bodied for their survival. Wannabes are entirely justified in believing they will be able to lead a productive, virtually normal existence after achieving their desired limb loss, although some no doubt underestimate the discomfort and difficulty involved in recuperation and rehabilitation. I myself attained four-year and post-graduate university degrees as an amputee, and since completing my studies have consistently earned salaries commensurate with my age and education. In this respect, my life is arguably no different than it would have been if I had retained my left leg, except for the added happiness its absence has brought me. - One author has suggested that having achieved her desired amputation, the wannabe is more or less doomed to disappointment and regret once curiosity has been satisfied and the novelty wears off. He proposes this as a justification the medical community could use to deny the desired surgery as therapy for the wannabe. Of course this argument is based on the two presumptions that curiosity is the sole or major motivation for seeking amputation, and that the surgeon's belief that amputation will ultimately make his patient unhappy is ample reason to withhold it. Dealing with the latter point first, the medical community could reasonably assume most people might be made unhappy by limb removal for any reason, yet obviously they do not let this stop them from performing amputations where its suitability is indicated by medical rather than psychological reasons. It is inconsistent to argue that the patient's potential unhappiness is sufficient reason not to perform the surgery in one situation but not in another. The true question here is whether the wannabe's real unhappiness caused by the unsatisfied wish for amputation is exceeded by the potential unhappiness which might (but might not) follow amputation. Physicians have no way of knowing the answer to that question, and are not justified in denying surgery for the sole reason that the patient might be disappointed later. Indeed, if that were sufficient reason for doctors to refuse to perform an operation, what elective procedures would be done? Any operation entails the possibility that the patient will be disappointed sometime in the future. If there is even a reasonable chance that amputation might provide effective relief of chronic unhappiness that is sufficient to impair the wannabe's daily function, it is logical to suggest that the medical community might be obligated under its own code of ethics to consider amputation as a viable treatment. If the incapacitation created by an obsession with obtaining amputation is greater than is normally associated with limb loss, surely amputation becomes the superior option. - It may be true that unsatisfied curiosity provides the sole motivation for some individuals' wish to be amputated. I have no reason to think it does not. Speaking for myself, however, it was only one of the three main aspects of my desire for amputation, as I have already described. It was the most intensely felt of the three, and probably the only one that could not be even partially satiated except by surgery, but I always knew that my curiosity about the sensations of amputation would die almost as quickly as my severed leg if I succeeded in getting what I wanted. If curiosity had been my only reason, I am sure I would have been able to keep my desire under control and avoid surgery. Even at the ripe old age of sixteen I knew better than to take on a lifelong liability for the sake of momentary gratification. No, for me the most important reason to have my leg amputated was a deeply felt need to triumph over the handicap, to feel the satisfaction of overcoming the challenge of a permanent source of difficulty. This has continued to be the case. I was totally accustomed to the feel of my stump within a few days after surgery, but the gratification arising from living a successful life on only one leg continues unabated more than a decade later. - I suspect that few wannabes whose yearning for amputation is strong enough for them to actually seek satisfaction base their desire entirely on curiosity. The desire to know what amputation feels like is certainly powerful, but the very act of thinking about this is apt to produce other motivations. - But I do not think it is a foregone conclusion that a realized wannabe must inevitably experience disappointment and regret even if curiosity is the primary motivation. I do not have extensive statistical data, but I have maintained extended correspondence with five other individuals who also managed to achieve the wannabe-amputee transformation as teenagers, and I think their experiences should be considered even if they are not provably representative. - Two of these amputees are women, and the other three are men. Two are in their twenties, one is thirty-six, and the other two are over forty. All three men are single upper-extremity amputees by the simple expedient of deliberately putting their arms in the way of farm or industrial machinery. Both women are above-knee single leg amputees. One obtained surgical amputation as I did, via referral to a surgeon by her analyst. The other claims to have paid an acquaintance three hundred dollars to destroy her knee with a shotgun. Whether that claim is completely true I have no way of knowing, but I have met this person while she was on crutches, and can affirm that her limb loss is genuine. I can also state that she showed me a newspaper clipping reporting her as the victim of a hunting accident. - The time spent by these people in planning and preparation varied from years to practically none at all. As might be expected, two of the men accomplished their "accidents" impulsively, without prolonged thought. They saw their opportunities, and took them. The third man operated a large hydraulic metal-cutting shear in a steel-fabrication plant for over two months while contemplating whether he should fulfill his lifelong dream, finally sticking his arm under the blade three days before he was due to begin his last year of high school, severing it just above the elbow. One woman underwent psychoanalysis for three years before her therapist offered her the opportunity for surgery in a hospital operating room. (I found that interesting, as my own amputation was performed in my surgeon's office to avoid scrutiny by the staff at the local hospital where he normally did major procedures.) The shotgun victim took the longest to attain her goal, continually planning and re-planning it from the time she was twelve until she was almost eighteen.

All five individuals told me that unbearable curiosity was their most compelling reason to give up their limbs, with two saying they had no other reason they were aware of. All five also said their amputations made them permanently happier, and that they would not do anything differently if given the opportunity to do so. They all expressed enduring satisfaction with their stumps and with their lives as amputees. - I think it more than likely this is a skewed sample, as I wouldn't expect to learn about or hear from realized wannabes who have discovered their amputations were mistakes through the channels available to me, but I believe it does indicate that ultimate disappointment with voluntary amputation is not inevitable or even necessarily common. It also suggests that long-term satisfaction can result even when the wannabe is mainly consumed by curiosity. - In fact, I would contend that the reasons a wannabe desires to be amputated are probably not as important after the goal is achieved as the intensity of that desire. When one's wish to have a stump escalates past a certain level, it is the wish itself which becomes the primary problem, a dominating, all-consuming obsession that is itself a disability. At the point when I began psychoanalysis, I couldn't think about anything but becoming an amputee or accomplish anything more than the simplest tasks. I was functioning far below my normal level, and could do nothing about it. Even my psychiatrist could not help me regain full control of my mental processes. In the end I was given my amputation not because I wanted it, but because that was the only way to stop me from wanting it. And it worked. Regardless of the disadvantages that accompanied the loss of my leg, it ended my obsession. I could not continue to devote all my thoughts to obtaining something I already had. My mind was at peace from the moment I woke up after surgery, and the relief that afforded me was so great it didn't matter that I could no longer do some things that had been easy before. I had traded a large handicap for a small one, and the bargain was good.

So my friend who stuck his hand into the hay chopper because he wanted to know if he would really still feel his fingers afterward was not acting out of mere curiosity. It was the difficulty of living with the intensity of his curiosity that really impelled him. He'd been wondering about this question for so long and so hard that his mind wouldn't let it loose. The only way to free himself from it was to free himself of the hand. Once he did that, it didn't matter what the answer to his question might be (yes, he still feels his fingers, thirty years later, and can even wiggle them), because he finally knew what it was and could go on to other things. The relatively severe disability of a missing hand seemed trivial to him, compared to the misery he had undergone prior to his "accident." There was no letdown, no sense of disappointment, nor could there be, because he experienced only improvements. He got exactly what he wanted, and was willing to tolerate the physical consequences because at least they were endurable. - It is logical to ask why curiosity about the experience and sensations of amputation should reach such levels. After all, one can be intensely curious about many things without becoming obsessive. What is special about this thing? - I think the answer to that is two-pronged. First, a desire to know exactly what an amputee feels can ONLY be satisfied by amputation. There is no satisfactory alternative. The amputee can try her best to describe her sensations, but they are peculiar to a physical configuration different than that of other human beings, and only someone with the same configuration can understand her description as she means it, and someone who has that configuration doesn't need her description. Second, and more important, the wannabe knows this curiosity CAN be satisfied. She can know exactly what sensations come streaming up from an amputation stump. All she needs is a primitive surgical procedure, and all her questions will be answered. It is this potential availability of the answers that leads her down the slippery slope. They are right there waiting for her, but always out of reach. It is like being on the brink of starving to death while a banquet lies waiting on the other side of an unbreakable glass wall. You might want to stop thinking about the food, but you can't help yourself. - This, as at least one other writer has pointed out, is where any analogy between amputation for wannabes and transsexual gender reassignment surgery breaks down. I have wondered what it would feel like to make love as a man, as I suppose countless other women have done. But such knowledge is unobtainable in the current stage of the medical arts. I could have a sex-change operation, but it would not give me the male experience. I would not be a man; I would be a woman with no vagina and an enlarged clitoris. I still could not learn what my husband feels when he rubs his penis back and forth in my body until his testicles discharge sperm and his prostate pumps semen into my vagina, for the simple reason that I would still have none of those specialized organs. Recognizing the futility of my curiosity, I am not tempted to pursue it. Since my curiosity is limited to the sensations of sexual intercourse and not to the other elements of male life which might attract a transsexual female, I am content to let it go unsatisfied and turn my attention to other matters. - That is not the situation with amputation, however. The wannabe is perfectly aware the answers he wants are available from a surgical procedure, and is not dissuaded by any sense that he cannot learn what he wants to know. To the contrary, his curiosity is encouraged by the knowledge that it can be perfectly satisfied, while at the same time it is frustrated by the difficulty and even danger of obtaining satisfaction. Is it any wonder he succumbs to mental gridlock? - The sensations I get from my stump might not be exactly what you feel in yours, but my stump is real, all that remains of a leg I no longer have. It is not simulated, Whatever I feel in my stump as I type these words is my brain's legitimate interpretation of the signals emanating from a severed and rearranged mass of atrophied muscles, blood vessels and nerves wrapped around a stub of bone. It matters not that your brain might experience the same stump differently, because I am still the genuine article. Once I succeeded in obtaining amputation I became an amputee, not some superficially altered cosmetic imitation of an amputee. - I am certain the irony of the transsexual-wannabe comparison escapes no thoughtful person. On the one hand we have the surgical removal of internal and external genitalia (and the breasts, in the case of a woman) and complete loss of reproductive function as part of a coordinated change in appearance that necessitates a drastic change in lifestyle and manner of dress in order to satisfy societal norms for the target gender, all without actually accomplishing the basic goal which is the purpose of the procedure. On the other hand we have the surgical removal of a single appendage with the only direct effect that of altering the way the patient walks, not even changing his appearance if he chooses to employ a prosthesis, while being entirely successful at producing the sensory and functional changes sought by the patient. Yet the first procedure is available to anyone who can passes the psychological screening and has the money to pay for it, while the second cannot usually be had for any price due to "ethical considerations" that may have nothing to do with the deepest psychological needs and qualifications of the person seeking it. - The attitudes surrounding other available elective procedures provide even greater contrast to the prevailing situation with amputation. As a woman I am entitled to have my breasts enlarged, reduced, reshaped or relocated as I desire, so long as I can muster the necessary funds and avoid acting like a raving lunatic. If have trouble pleasing my husband I can have my vagina tightened. I can also have myself sterilized, replace my natural teeth with crowns, get my face lifted, have my nose bobbed and my jaw line altered, and so forth. I can even, believe it or not, have my labia retailored to better match my idea of what they ought to look like (even though I have somehow managed to struggle through twenty-eight years of life without ever seeing them or feeling a desire to do so). Some of these procedures could have serious effects on my life while others are intended only to alter my appearance in ways I hope to find pleasing, but what all have in common is availability on demand. If I can pay for them, I can have them. The medical community's position on this is, it is my body, and I am entitled to alter it in accordance with my notions of how it ought to look. - But somehow that position changes abruptly when the requested alteration affects my extremities. I am not permitted to seek alternatives to four limbs and twenty digits, even to relieve unbearable psychological pressures. Perhaps I am aware of receiving no discernible benefit from the smallest toe of my right foot, but if I ask to have that toe taken off in the belief it will improve the looks of my foot and make me feel a lot better about myself, I will be told that is an irrational request, that I need my toe even if I don't see that need. - Think about this. The same surgeon who tells me that the desire to lose a toe I will never miss is a sign of mental instability might well stand ready to change the shape of genital labia on request. Am I really less well-balanced than my sister who feels a need to alter the appearance of her little pink *coochie*? Maybe Sis and I are both disturbed, but the real irrationality here is in a system that caters to the disturbance of one, but not the other. - It is quite obvious that my desire to shed an entire leg is a different matter than someone's wish to have a minor toe clipped or a nose bobbed, but the difference is one of degree rather than basic principle. It IS my body. Were I convinced I would be a happier woman if only I could have a four-toed foot to fondle and gaze upon, who is to say I would be wrong with any pretense of genuine authority? Few sensible people would argue I would harm anyone but myself, or that the risk of that is more than trivial--people live their whole lives with missing small toes without being conscious of any meaningful loss. Going further, if I know about and am willing to accept such things as phantom pain, physical rehabilitation and a lifetime of inconvenience in order to obtain what I perceive as the benefits of having only one leg, who empowered my doctor to deny me what I want, in light of the many other things he is prepared to do at my request, some of which also have negative side effects and entail permanent alteration of function? - Which brings us back to sex-change surgery. The transsexual patient gives up functional genuine sex organs of one kind in exchange for something that vaguely resembles the other kind but doesn't work at all except in the most rudimentary sense of providing something which may be inserted in penis-like fashion or which will accept insertion of a penis. The simulated genitalia will most certainly not provide their owner with sensations anything like those routinely experienced by people who were born with the real thing, and he or she must be satisfied with whatever he or she gets. The necessary operations entail major discomfort and lengthy recuperation. Yet sex-change surgery is considered worthwhile medical treatment because it provides the physical appearance and semblance of function for which the transsexual has endured a lifetime of yearning at obsessional levels, thereby alleviating great mental torment and improving his or her overall emotional well-being.

I submit that when all facts are considered, if there is any difference in which procedure ought to be accepted, prescribed and performed, it ought to favor the wannabe-amputee transformation over sex-change surgery. In contrast to the latter, it provides exactly what the patient wants, not something that merely looks like it, and is therefore a more successful and effective procedure at the physical level. (To be as successful and effective, transsexual surgery would have to give me a complete and functioning set of male organs.) Leg amputation involves relatively little post-surgical pain compared to other major procedures, as well as brief hospitalization and relatively short recuperation. Removal of all or part of an arm imposes even less stress on the recipient's system, with patients often up and walking around only a few hours after the surgery is performed. Amputation is far more cost-effective than gender reassignment, with the total fees for above-knee leg removal being less than five thousand dollars even in an expensive American hospital, a small fraction of the price for a sex change. (The economics might not be quite so favorable in the case of surgery involving the torso, i.e., *hemipelvectomy* or forequarter amputation, but I have yet to hear of a wannabe desiring such a radical procedure, and cannot imagine one doing so.) As a significant downside, amputation frequently entails phantom pain of varying duration. I doubt many wannabes with a serious interest in pursuing surgery would regard that as other than a reasonable risk, however. - More to the point, amputation can work very effectively as a treatment for the wannabe's obsession. I can attest to this from my own experience as well as that of the aforementioned amputees with whom I have contact. There are also two cases of older men obtaining amputation through self-inflicted shotgun wounds which are well-known in the wannabe-devotee world, and these also experienced favorable results according to published accounts. Such anecdotal evidence does not constitute adequate grounds for doctors to offer a procedure as treatment for any condition, of course, but it should point the way to a more rigorous evaluation under controlled conditions, where the surgery is performed on carefully screened candidates. - Note that I am not suggesting sex-change surgery be abandoned. To the contrary. It continues to be performed because it has shown itself an effective means of bringing transsexuals a measure of tranquility and contentment, thereby improving the overall quality of their lives. Rather, I propose that the mental torment undergone by wannabes be recognized as a seriously debilitating condition similar in nature to and as important as *transsexualism*, and that amputation not be ruled out as a reasonable way to treat it, just as gender reassignment is used to treat transsexuals. If this is done, I have every confidence that wannabes who are likely to benefit from amputation will be properly identified in psychological testing, and that amputation will come to be recognized as a valid therapy for the condition. Should that happen, wannabes will be able to come forward with some hope of finding the contentment they seek, and I believe we will find this is a much more common condition than has heretofore been thought. The alternative, of course, is to continue the status quo. In that event, I would contend the medical community is guilty of maintaining serious inconsistencies in the way it approaches elective procedures, and of turning a blind eye to the overall well-being of an admittedly small part of the population who now have no legal, safe and effective means of treatment available to them, apparently for no better purpose than the propping up an outdated view of what constitutes necessary surgery. Then, as now, wannabes will be fully entitled to wonder why other major surgeries with no more serious purpose than to confer a desired change in appearance are available on request while our deepest desires and disabling frustrations are regarded as frivolous at best. Then, as now, we will ask the medical community, "Whose body is this, anyway?" –

After my discussion about the middle aged man, and the fact that there are so many female wannabes I can say that the biggest group of wannabes are maybe males, but for sure not in the middle aged category. Let us also not forget, coming to the subject females, that the most woman on this world are not such a easy talkers as the most man. An example of this is the internet. Ask a man for details and also a woman. The man will tell more about him then the woman. - The wannabe woman I talked with in the past never were open to conversations from the first moment. The conversation and the trust needed to grow before they really mentioned the fact that they are wannabes. - Also we may not forget that the first real contacts about the wannabe subjects and between wannabes were on the internet. Coming to the point of internet and internet connections we may not forget that the basic language on internet is English and also that the best connections are in Europe and USA. The countries with a lot of white people. Now coming to the point of black people and being a wannabe we can ask ourselves why the African wannabes are not with so much. Or are they?, but maybe their English is very bad and maybe even their internet connection is not that good. Even on the first Ampulove site I had for years, it was very clear in the visitor statistics that the biggest group of them came from North America and Europe. I never had any email from someone from Congo, or one of the other African countries. In one of my travels I did to Brazil and even being in contact with Brazilian wannabes I discovered that for the most of them the internet wasn't that easy to get in contact with other wannabes, to talk about their feelings, just because their English isn't that good. We may not forget that In countries like Brazil the most people need to pay a lot for English classes. So the most Brazilians don't speak English.... Wannabe and Amputee sites are mostly in English, while Brazilians speak Portuguese. So, very difficult to get in contact with the wannabe world, other wannabes or even with psychiatrics and psychologists that do investigations on them.

To end up the subject *black apotemnophiles*, and or they are with a lot; Let us wait a few years till the internet is worldwide much better, till the English language of the most wannabe sites offer Multilanguage's.

Why I still use the name apotemnophile and wannabe is because I disagree with a lot of points that the original BIID term explains.

The fact that it is not sexual, what it is for the most of (successful) wannabes; The fact that the biggest group discovered their wanted feelings on a very young age.

Wannabe, want to be, BIID. Let me split the group wannabes in two parts.

There is a big group of wannabes that never really will become an amputee. People that even don't think about becoming an amputee. The two groups I make are: Want to be and Need to be. The group of want to be don't need the amputation, are basically pretenders and even doubt about becoming a real amputee. They know and understand that once they are an amputee they can not turn the clock back in time. Basically it stays with fantasying and acting as an amputee. The other group, what I call need to be is a group of people that will do everything to become an amputee. Some of them even dangerous things. A few need to be people cut of their own small body parts; Toes, fingers. Trying to become an amputee by using dry ice, local anesthetic injections, faking accidents that sometimes end up very bad for them.

Years ago I had a very good friend on internet who was a wannabe and wanted a leg off. He died after trying to have his left leg off by laying down on the rails, with the hope that a train would cut his leg off. But the train took his whole body up and he died. Something he didn't want but thanks to the ignorance of surgeons he visited before with the scream to help him he did.

That wannabes of the group that really need the amputation done, and need to end up in such a sad way, is a tragedy when you think about the group of transsexuals on this world. When a man want to become a woman or a woman want to become a man, at those days it is simple to get the difficult surgery done.

After the transgendered surgery they still need to take their whole lives hormone pills.

Someone that want to become an amputee goes through a simple surgery of about an hour, for a leg amputation, don't need to take a whole life long hormone or whatever pills.

But surgeons don't agree with that and like a big group of the medical world, they see a wannabe as someone who want an handicap.

Silly when you think about that a wannabe don't see their selves as a handicap person after the amputation is done. The wannabe group that is still apotemnophile and need to have the amputation done, consider themselves as handicapped in a four limbed body. Only the amputation(s) will let them feel completely and will help them to get rid of the bad, sometimes depressing feeling they have, living with all the limbs.

Surgeons as Dr. Smith from Scotland did, are hard to find those days. For the most wannabes it is then very difficult to find a surgeon that agree with doing the amputation(s).

Since a few years some surgeons from countries like the Philippines, Russia, Sri Lanka and Mexico; -just a few countries to mention, do amputations on wannabes. Through different communities and internet groups the wannabe can come in contact with a surgeon from a far country to have his or her amputation done. Mostly for an expensive price. We are talking here about $10.000 for a limb, in the cheapest ways about $5000. For this price, the patient receives the wanted amputation. We may not forget that the amputations in those countries are not always done in a safe way and can be a danger for the person that goes through such a surgery.

A good example for this was my travel to Sri Lanka a few years ago, where I met a German psychologist that lives there. He wasn't a wannabe or devotee but was very interested in making money out of wannabes. He asked me for lists and addresses of apotemnophiles that I know, but for sure I wasn't interested in giving him any information about people that contacted me during all the years. One night, sitting in front of his house, we had a conversation about his idea about helping wannabes. I asked him what he would do if something would go wrong during the surgery and the wannabe would die.

His answer? ; ´We will burn the dead body, no proofs will be left behind´. So every wannabe that paid first a huge amount to become an amputee, that could die in a surgery, would be burned up afterwards, if something would go wrong. Dead, no one of the family that ever would find out what happened with their family member.

Near that, this kind of surgeries in foreign far countries are not only a risk for the wannabe and expensive, but are in my eyes only a possibility for rich people. Probably not every wannabe have the $10.000 ready to have a leg removed.

After going deeper into this subject with a good French wannabe friend of me, who is also in contact with a lot of wannabes, he told me that much more strange things are going on sometimes in the direction of surgeries in foreign countries.

Some surgeons who do amputations on wannabes for a lot of money are a kind of freaks. A story from Morocco from about sixteen years ago;

A guy went there to have a leg amputation done, paid the surgeon well, but when he waked up, he needed to discover that more was amputated then he had asked. The man was off course afraid afterwards to start a case against this surgeon. How would he tell in front of a judge: 'I wanted my leg off and paid that surgeon'...

In the world are some more freaky surgeons that only thinks on money and not on the health conditions of their patients.

In 1999 there was the story that in San Diego, a former doctor, John Brown, faced life in prison after being convicted of murdering a man who died of gangrene after paying $ 10,000 to have one of his healthy legs amputated. The dead man, Philip Bondy, 79, was a apotemnophile and had the desire to have a leg removed. Brown, 77, lost his medical license in 1977 after botching three sex-change operations he performed in garages and hotels. In Miami, police searched for Reinaldo Silvestre, 58, who pretended to be a plastic surgeon and operated on an unknown number of patients using animal anesthetic. In one case, Silvestre gave women's breast implants to a male bodybuilder. And in Brazil, authorities considered pressing charges against Dr. Alberto Rondon for marring more than 140 patients over the past 15 years while performing plastic surgery without the necessary training.

Sites like BME, *Body modification Ezine,* has a shop online were people can buy local anesthetics, simply everything to do *home surgery*. Or this is allowed is a good question, but for sure it is not safe to do all those things at home.

In all the do-it-yourself stories I heard in my life, the most wannabes became an amputee with using dry ice. Dry ice is the solid form of *carbon dioxide* and have temperatures low as −56.4 °C / −69.5 °F. This give burn wounds as result, when you hold it to the body, and can damage in a very hard way if used for a few hours. Gangrene is the result.

The wannabes I know that used dry ice didn't became all an amputee. A lot of them went trough a hell of pain and knowing that surgeons do everything to safe a limb, only a few reached their goal.

A friend from Germany, who wanted to become an amputee, became a successful wannabe after using dry ice in a liquid form. He told me that he suffered almost two full weeks from terrible pains in the body parts that he didn't damaged enough, before surgeons even consider to do the amputation. But with a lot of pains he finally had the amputation he wanted.

Some surgeons are for sure open minded to wannabes;

A friend who was a below the knee amputee and wanted to have an above the knee amputation done, visited only tree different surgeons to have it done. The only thing he told to the surgeons was a fake complain about a pain he didn't had. He told the surgeons that he was in a terrible pain in his stump and knee, and that he was better of with an above the knee amputation. One of the tree surgeons believed in his words and helped him in becoming an above the knee amputee. He never told the surgeon about his wannabe feelings, but considering how easy he got it done, we maybe can think about the fact that the surgeon knows about the existence of wannabes and helped the guy, just to not see that he would do dangerous things to himself afterwards.

A French wannabe friend, let me call him Ludovic, was for many years in contact with me.

Special was that he is a doctor and wanted his left leg amputated above the knee. But even being a doctor himself, he found it to difficult to reach the above the knee amputation he wanted so badly. Asking his doctor friends with who he worked daily to have this surgery done, was impossible. Never before he had talked to anyone about his wannabe feelings. The first time he did, was to me and on internet.

Ludovic told me that he had a machine that could find the main nerve in a limb and then would inject local anesthetic in his leg. He prepared a fake accident, rented a motorcycle and went on vacation somewhere in the South of France were two wannabe friends were with him. At night he went to a place where there was almost no traffic around that late hour, somewhere in the hills behind the city of Marseille. A perfect place to have his accident done.

Once his leg was anesthetized, he used an hammer to damage his foot so badly, that later in the hospital surgeons only could go over to amputation.

He damaged the motorcycle and told later to the doctors that he was in a motorcycle accident and that the motorcycle smashed his foot.

This are only a few stories from a few people. A few people that were wannabe and became an amputee by wish.

A long time ago I received once a very long email from an American woman that wrote me her complete life down.

She told me that she was a judge, that she was happy married and that she had children. She explained me that she was so happy that she had founded me on the internet, because she wanted to talk already for many years to me. The subject was indeed amputation. Years before she was in a car accident in the North of the USA, and stuck by snow it took hours before someone even came to help her. That time, in the freezing cold, her legs were without feelings from the heavy cold. She didn't mind about it and even from the leg she wanted to have amputated, she removed the shoe, so the freezing cold could do better his job. Afterwards, surgeons needed to amputate her leg above the knee. She never told anyone about the fact that this real accident only helped in her way to become an amputee.

What first was an accident, became for her a solution to have her amputation done.

Basically a wannabe knows what he or she want. It can be a single above the knee amputation, it can be a finger, but it can be even a double above the elbow amputation.

That some wannabes really want their two arms amputated is even for some other wannabes a big question mark, why the hell someone would be more happier without arms.

Isn't that a little bit to much at once?

In my group of friendships with wannabes there was an American guy from about twenty four years old that absolutely wanted to become a quadruple amputee. Someone without legs and without arms.

His big dream was to find afterwards a girlfriend that would take care of him and would help him in the same way as a newborn baby need help with everything.

Myself, I don't consider this as being a wannabe in the direction of apotemnophilia or even BIID. More, I consider it as a kind of fetish for being helpless and the 'take care of me feeling'. Some other fetishes are in that way that people like the take care feeling. A good example is the diaper fetish. Adult people that wear a diaper, acting as a baby and an other person that take care of them.

The guy who wanted to become a legless and armless amputee, didn't became it yet. A few times he asked me to meet him in person, but every time when it came to meeting each other for real, he canceled. *Funny* part of the whole thing is that he removed landmines in Angola, that was his job. He could had jumped on a landmine if he would like to become an amputee, what he even told me he would do.

This kind of people are fantasying a lot about being different. Maybe it is just a scream for attention.

The group of wannabes I know, that really became an amputee by wish all confirmed me the same thing; They didn't regret that they became an amputee, but they regretted that they waited so long before they became it. So, we can consider that the wannabes who belong to the group need to be, are really happy with their amputation(s).

Years ago, when I met the people involved in the investigation for the *Horizon* documentary from the BBC, I remember that one of them told me that a wannabe knows exactly what amputation they want and that the feeling will stop after they get it.

Let me say that this is wrong.

Even one of them, who became an amputee in the hospital in Scotland with the help of Dr. Smith called me afterwards and informed me that he was thinking on having an arm removed too. He found a girlfriend that wanted to become an amputee like him. This friend wrote me several times. He was also one of the first persons that mentioned that I should need to write a book about the apotemnophilia and acrotomophilia phenomena's. In one of his letters, he mentioned also how happy he was with his wanted amputation. (See the next two pages – a print of one of his letters to me that I received)

People that become an amputee, starting with a leg, because they wanted that one leg off and end up with a double above the knee amputation or even more, is for sure not so strange. There are more of those cases. A few wannabes told me that they discovered after having the amputation done, that the feeling in their stump was so good and a kind of warm erotic oriented sense, that they wanted more of this sort of feeling. The only way to got more of that feeling was having more stumps.

Not only in the wannabe group we find this kind of people back. Even in the amputee world I know a few amputees that became a wannabe afterwards. Someone in an accident, can discover afterwards that they like it so much to be an amputee. A good example of that is a friend from the Netherlands. Years ago I went out with him and a woman from Belgium, who was a devotee. So, a female devotee, a guy from the Netherlands without legs, and me. Having together a beer and talking about amputees, devotees and wannabes. He told us that he was a prosthetic maker and that he was in an accident before in the USA. He lost his two legs above the knee after his car hit a truck. After the surgery and being an amputee he discovered about the wannabe world. Years later he contacted me by phone and asked me or I didn't know a surgeon that could

F. ███████, ████████████████████████, Germany

January 2007

Hey Alex,

A friend gave me your address and I hope you are dont mind if I pick up contact to you this way.

██

I envy you for lots of courage! Hopefully you are finde, doing well, and getting around in a fair manner, whatever is left to be possible.

Its been more than 6 Years already since I had the leg off. You remember? A middle sized right above Knee! I cant complain. N₀t at all. I am happy, satifaction/satisfied and having the usual problems especially with the artificial leg. And of course with doctors/prothetists/ladies ....

No, the insurance made in the beginning trouble, They cut off some of the paxments, but now they gave their OK to the 2nd new C-leg.

I think I am not emotional the type of person who likes to wear this still clumsy appliance. Without I like to be mostly whenever I am not forced to wear the leg.

I think you know, that they cut my leg off twice. Some sort of a bone-spur on the distal end caused some trouble. I felt my stump was to long and so they shortened it about 2 inches. I rather would have had that they woul have it cut off shorter yet.

A friend of mine finally got it all done at once. Right above knee, left or right above elbow, and some fingers. I guess he is gay and appears to be very happy - no searching for solutions any more, no pretending, no longing, no thoughts of what it will be - he has got it!

I lost contact to him mostly becaue I lost his address and stopt writing. Too bad! I loved his honesty and the way he wrote letters.

From what I heard, is that ███████ move to Holland, had one and now both legs off.

I talked to a guy who had the right arm off some time ago. He worked with computers, is married, middle aged and as he told me his wife was not informed. The whole proceedure took only halve an hour, but he had to stay in the hospital another couple of days.

In the beginning the stump was uncontrolled shaking, some sort of an involuntary shivering like sensation similar to freezing when in the cold. He now is fine and enjoys life! Even his wife can cope with it after long discussions of why and nottelling her about it.

Well, it is not everybodies baby! But why suffer?

What does it matter? one or two more amputees? 60.000 are losing a limb as cause of smoking and similar sicknesses.

I learned that good, real friends become more and more rare. Everybody wants, few, very few go through the ordeal.

I would be keen to hear from you again. Tell me about you eyperiences, how you get around, how is writing left handed ? and whatever you want to take (write)about. Can you use crutches with only wearing one leg? Is it easier to get used to the RK prosthesis?

Financially things are getting hopefully better here in Germany. I dont know about Belgium. Anyway, I sincerely hope that you are fine, of good health and of course on the way to become well to do?

My biggest problem in a situation like you are, would be the gaining of weight. Me too as it is now, am lacking energetic motions/movement

Letter from a friend – see page 57

burning up all the calories.

I dont think I would like to wear an artificial arm but it would be interesting. It is funny, but from the little I know, I would like to prefer a fairly short BK and fefinitely nothing longer than 8 inches.

Most ladies I met, also amputees, are afraid to talk about this subject. The few exemptions are rare and in good hands!

I have to do with real estates. Mostly developemants of old sites. Small industrial sites have to be converted and modernized. A tough hard business. Lots of competiton, prices incline constantly, engaging companies from off-road, from low-priced countries such as Portugal cannot be avoided. Here the laws forbid,or I better say, allow only in certain cases such employements.

I talked to someone from Belgium who legally brings his workers over here and writes bills to the german contractors from Belgium. Also not only this is an advantage, but just as well, that taxes are lower in your country as I am told.

Alex, why dont you write your story? You can change names and dates. I think it would be, if written accordingly, a well sold publication. It would give to many interestes parties, first hand information.

Not only to poeple who want to be, but also to professionals dealing with this subject and of course broad minded individuals who search for information of this kind.

I think, but you know it better, there are plenty of men and women who like to know about it. You are predestinated to do so!

I would offer my assitance if you like.

By the way: where is Ostduinkerke located in Belgium? How is your Family?

Alex, I wish you all the best for the New Year 2007 and would appreciate to hear from you.

                                  Best regards,

pls. write on my address:

If you like, give me your phone number!

Letter from a friend – see page 57

help him to become an above the elbow amputee. He told me that he discovered that he enjoyed being an amputee so much, that he wanted more amputations. Or he really became an arm amputee or not, we lost contact afterwards.

But, a good example to show that some regular amputees become a wannabe afterwards.

Why this is, till now there is no very clear answer on it. Maybe we can go deeper in on the fact that someone who became an amputee get more attention, that they discover how things can be done in a different way; A way that a lot of people give them an extra attention. A lot of those examples are on YouTube. A site filled with millions of movies and also movies about amputees that shows how they can do things on one leg, or even without legs. A few examples are people that climb mountains, even someone that reached on one leg, is from Belgium, the Mount Everest, the highest mountain on the world. All those extra attentions that some amputees get, can make them more happier being an amputee then they ever were happy in an earlier live, when they weren't an amputee. This extra attention those amputees get, can give them maybe the feeling of –wanting more. More amputations done, more attention?.

Or the biggest group of wannabes will be Africans, black people, woman... We have no any idea about this at the moment. For sure the Germans are at this moment with the most. From a lot of site statistics on internet we can make that up.

Why Germans? Has it something to do with the war of Hitler and the fact there were a lot of war amputees? Or are Germans more open minded?

Or it has something to do with Adolf Hitler's war, I don't think so. My question is then why the American civil war, that also brought a lot of amputees, or even the Iraq and Iran wars didn't made more wannabes. Both wars had as a result that a lot of wounded soldiers came back home with an amputation. Comparing with Germany who has the most talking wannabes, then we are surprised that not the USA have the most. USA is totally wheelchair accessible, while Europe for sure isn't that kind of paradise for wheelchair users.

That could be one thing, that wannabes would be with the most, in a country that is very well wheelchair accessible, but it isn't.

The other point around this case is that with my investigations on wannabes, I discovered that the real wannabe, who is in my eyes a need to be, don't think about life afterwards, in the way, what will happen after amputation. Or they will still be able to work, to walk, to play football or whatever they do in their non-amputated life, the biggest group of need to be´s don't think about this. Another theory comes from Professor Anderson from the University of Georgia, 7 years ago:

-The old souls who come back in a new body and who are in the need to recover their old body. Germany and France have the higher number of Amputated guys , with their terrible wars, in the past...

Thinking about live after amputation, the most wannabes that I know, want a limb off, but don't want to replace it afterwards with a prosthetic. A wannabe told me once: ´If I have my leg off, why should I replace it afterwards with a fake leg, then I better can keep my real leg´.

Maybe the German people are just a little bit more open minded and maybe they published more on television and magazines about the fact that there exist wannabes.

How more people know that wannabes exist, how more who are a wannabe will start to talk about it.

So many times wannabes told me that they were happy at the moment they heard about other wannabes, because for them that was the moment they finally realized that he or she wasn't alone with that feeling of *wannabeisme*.

So many of them said: ´The day I heard I wasn't the only wannabe on the world, I felt me much better´.

So many wannabes thought that they were the only person that had this desire to become an amputee. If someone that don't know that there are other people on earth with the same feeling, it is very difficult to start talking about a subject that no one ever heard about before, including yourself.

Who always was an example for the wannabe world was an American named George Boyer. He was a wannabe and became an amputee after shooting a bullet in his left knee.

Since the beginning I was always in contact with George and he liked very much the Ampulove site I had at that time. He wrote me a few times by email, but also letters by regular mail. George was a very open minded wannabe and very old. Even on the age he became an amputee, he wasn't that young. George was in the documentary ´Whole´, that I mentioned before.

For me he was an example for the wannabe world. He became very open minded about it, in a time that it was all big taboo. He was in different magazines and told open that he shoot his left leg off. George lived in Orlando; Florida, USA. When I moved years ago from Belgium, to live in the United States, my idea was to visit George. I tried to phone him, at the number he gave me before, but unfortunately I needed to discover that George died. George dead gave me the inspiration to go on being open minded for the wannabe and devotee community on this world.

About the amount of wannabes on this world, one on thirty thousand, I can say that there are probably much more wannabes then this. But looking at the fact that this are old statistics from a time that internet wasn't so big as it is now, we maybe need to think that there are much more of them. Not everyone dear to talk about it, not every wannabe knows that their feeling have a name. Near that a big group of apotemnophiles will not come out for their feelings. Their feelings will stay a secret because they are afraid to talk about it, or ashamed, or just don't know there are more wannabes on the world.

Wannabes have basically a normal life, even after amputation. It is for sure not a mentally illness that is written on the face, or visible. With other words, a wannabe just runs a normal life, goes out, have children, is in business, have work, hobbies... only that one feeling is different with other people. They want amputation(s).

A few good examples of normal active wannabes I know that have a normal live;

I mentioned already the female judge from the USA and a French doctor. But there are much more examples of apotemnophiles who have just a normal live.

A French businessman who travels around the world, A professor from Switzerland who give lessons on different universities worldwide, an American professional dancer, a German fireman, a Belgian guy that works for the tax control, a police officer and an immigration officer from the United States, a French actor,...

All real wannabes or people that became on request an amputee. All people that have a normal live.

A few of them I met, others I know from the virtual world.

Years ago, in the times that people didn't know about this subject, a few wannabes came in contact with psychiatric hospitals. A good wannabe friend of me was so foolish to inform everyone he know about his wannabe feelings. He told it to his wife, to his best friends, even to a doctor who was a friend of him.

This all with the hope that they would understand and accept him and his wannabe feelings.

The result? His friend and doctor that never heard about wannabes placed him in a psychiatric hospital. A lawyer helped him out of it after five days. Afterwards the whole city was off course informed about him as a wannabe. People were laughing with him and also he lost his wife and all his friends. People that told him in the first place that they understand him, that they accepted him as a wannabe.

Years ago transsexuals were basically locked up in the psychiatry; Same with wannabes. But now, years later we all know that transsexuals don't go anymore to the *shrunk-* hospitals, but get help in a different way; Becoming what they want to become. If not a woman, then a man.

Same kind of situation is going on with the wannabes. Years ago when no one ever heard about wannabes they first would be *house vested* in a psychiatric hospital. Now those days, more and more psychiatrist knows about the existence of apotemnophiles and try to help them in a different way. But just like it is for transsexuals, there is no any medication that can stop the wannabe feelings. The only remedy that helps in this way is doing the amputation(s).

If we look at the transsexual community, then we know that they find those days easier help to get it done, then years ago. When I called once with a psychologist, the woman confirmed me that they still don't know correctly from where the transsexual origin is coming from. The same with wannabes. Where it all really comes from, the origin, is very difficult to answer. The only thing we know is that they exist and that they also need help, just like transgendered people.

One thing a lot of apotemnophiles will agree with, is that the feeling is there from the beginning. A thing you are born with, but that grows with the years.

You don't become a transsexual, you don't become gay, but you discover it in a later age stage when you get conflicted with someone or something that gives you a sign. In this way a sign like: `Why I am not an amputee?´.

Talking about the signs by wannabes, I had also a few conversations of that during the last years with some of them.

A few reactions of people I remember; - My uncle was an amputee, I wanted to be like him; -A girl in my class didn't came to school one day. The teacher informed us that she was in the hospital and that surgeons needed to remove her left leg above the knee, due to bone cancer. Since that day I wanted to be like her.

The wannabe feeling grows by the years. In the most cases I know it started very young. In a normal way that the wannabe isn't busy with it the whole or everyday, but how older he or she get, how more busy they are with it.

Eventually the feeling will be so strong that the biggest group of wannabes will be focused every day on the subject amputation.

I know wannabes that went to libraries, for information about how people become an amputee; Wannabes that even go to rehabs to see amputees, just with the hope to see someone like they want to be; Apotemnophiles that studied the medical direction, because they wanted to be in contact with amputation.

Here we are coming to the next point that a lot of wannabes are in the medical world. Surgeons, doctors, nurses….

If you are interested in the world of stumps, then the best place to work in Is for sure the medical world.

Years ago I was in contact with an amputee through the internet. The poor man never heard about wannabes. When I told him that wannabes are people who wanted to be like him and me, he was openmouthed that this really was possible. He could not imagine that even someone could be jealous on his body. He had an above the knee amputation.

He was very depressed about it. More and more he went into details and told me about his life, his wife and children. He told me about the surgeon that had done his leg amputation. More we talked about the surgeon, more I realized that I probably knew this man. After asking his name I was the one who was surprised. The surgeon who had done the amputation was a wannabe and amputee himself. A lot of people in the wannabe world knew him. The surgeon became an amputee himself but died a few years ago.

About amputation by wish we need to know that there are more stories and facts then only the real wannabe cases. A good example is Lula, the ex-president from Brazil. Some stories in Brazil going around that he became an amputee by wish. Before he was a president, he lost his index finger in a work accident. Some people saying that he lost his finger on purpose to get money from the company.

Years ago I was in contact with a German guy who wanted his leg off, but before he would go on with that, he was planning to have a good insurance. An insurance that would cover an accident that he was planning. In that way he could receive after the amputation a bunch of money.

Let us go back to history and war. How many soldiers didn't cut of a finger, so that they weren't able to pull the trigger of a gun. In this way, by becoming an amputee he didn't need to go to the war.

Since a few years only, we know that it is possible in some countries to have toes removed. Not for the same way why a wannabe want to become an amputee, but to be able to wear nicer shoes, shoes with a smaller tip.

Near toes are a second group of small body parts; The fingers. A few years ago I peck up stories about places, basically in Germany and mostly tattoo studios, where people could go to have a finger or a toe removed. Not because they are a wannabe, but because they were into body modification, and found that having a finger or toe off would look *kinky* on them. This all belongs more to the world of body modifications that can go very far.

Earlier in this book I mentioned the website BME. Thinking on how many people are into modifications, looking at their statistics and having over a million of pictures online only about modifications, then we can conclude that for sure a lot of people are into different modifications.

Let us go deeper in on the subject body modifications and apotemnophiles.

During all the years I was in contact sometimes with some people that told me that they are wannabe, but near being a wannabe they wanted more. Some of them inject silicone in their penis to have a bigger one, some of them had a penis amputation. But, the most of them were into more sexual aspects. Latex, trio, group, SM and more...

The link between this all is that we all can discover something we like or going to find interesting. But the need of walking around in latex clothes or to have group sex, is not in the same way as a wannabe that has a need for amputation. The need for amputation is an obsession that keeps a wannabe the whole day busy, till he get his wanted limb(s) amputated.

In this way we can consider that some people who are into different sexual aspects find amputation an interesting thing. They find it basically a *windup*.

The most people interested in little body modifications or different sexual unusual things, knows that what they do, just like wearing latex clothes, are temporary. Even if it comes to their gender modifications; penis amputation, nipple amputation, toe or finger amputation, they will not be irritated in walking or daily life activities.

When we go back to the group of wannabes, then we know that their life will change after amputation. A wannabe who want to have a leg off, will need to have a prosthetic, crutches or a wheelchair. This is not the case with body modifications. Body modifications don't belong to the group of wannabes, and are basically pure sexual and temporary oriented.

Having a leg or arm removed, is for life. Having your nipple or big toe off is for life too, but will not bring any drastic changements.

Then we can talk about the group of plaster pretenders. A plaster cast on your leg, your arm. So many people like to be in a cast, like the warm feeling of a plaster on their leg; The attention from people saying: ´How did you broke your leg?, sorry to ear that´. Somewhere we can connect this to the world of apotemnophiles.

Personally I know some wannabes who were in a plaster cast many times. Their explanation is because the cast give them a kind of –being an amputee feeling.

Basically the feeling of a real amputation is that you still feel the amputated limb, but that you cant move it and that it feels warm or like it lay in a nice warm bath.

Going to look deeper into a plaster cast. Being in a plaster, you cant move the limb, it feels warm.

A lot of plaster cast pretenders do the same as amputee pretenders. Putting a cast on, using crutches, going outside in public.

Also in this case of plaster pretenders, we put them in the group of temporary.

A plaster can be taken off. An amputation stay.

On the internet we can find a lot of people that have a plaster cast site where pretenders of plaster can find information about how to place a plaster in the correct way.

Some of them have an almost private doctors room at home, where they put plasters on people who want that.

The biggest group of wannabes is for sure amputation oriented. Near that there is a group of people that would do everything to get paralyzed, some people that dream about: ´never being able to walk again´. Some of them, just like with wannabes, they have a wheelchair at home.

Why someone is interested in the subject –never walking again is not sure. But some movies, images on the street or in daily live can play a roll in it. Seeing people who cant walk, having a happy life, but are wheelchair bound, can give the wannabe an impulse in the direction of a life, that even without walking, can go on in a normal way.

If you ask me now what the most strangest story was I ever heard about wannabes. For sure an American young woman who contacted me trough internet a few years ago. She told me that she had her two legs off high above the knee, from being in a car accident. She was a wannabe before, but not an amputee wannabe. She wanted to be paralyzed. After waking up and seeing herself as a legless girl, she liked it so much that she wanted to become blind on one eye, she wanted her nipples removed, her clitoris and wanted to be deaf. I told her that I

75

didn't believe her, that everyone can tell me this kind of stories. So I gave her my phone number, asked her to call me, what she did. On the phone, she confirmed me everything what she had wrote down before.

She told me that she wanted to visit me, that she would like to be in love with a legless man. But afterwards, days later, back on internet, she purposed me to meet her, to become together blind and deaf. To find us someone that would take care of us.

´Very weird´, was my conclusion.

She send me a picture of her. Naked, laying down on the seat. Showing her stumps. So short, almost like a double hip amputation. The picture was so little that later I zoomed in on the stumps and discovered that it was well done *Electronic Surgery*.

Electronic Surgery, or in short terms called ES, is a kind of picture modification. A lot of wannabes like to see how they would look when they have an amputation. Taking your own picture, removing the legs or arms with some special software like *paint shop*, and you can see yourself as an amputee. During all the years the amputee and wannabe communities on internet have been grown, there are a lot of pictures from people who are original four limbed, but have on the picture an amputation. Some wannabes are real artists in that. There exist pictures on the internet, that it is very difficult to see or it is from a real amputee or not. Years ago, some wannabes started with making those ES pictures from themselves, other devotees started with it because there weren't enough real amputee pictures on the net.

From wannabe to devotee. Some wannabes are also devotees or contra verse.

There are wannabes who don't only want to become an amputee but are interested in being together with an other amputee. There is indeed a big link between wannabes and devotees. Both of them are interested in stumps, one want the stumps, the other likes stumps.

A wannabe who is also a devotee isn't that strange. If someone want to have an amputation, it all start by looking first to amputees. When you see an amputee first, and then want to be like him or her, then the link is easy understand when we try to find out why there are wannabes who are also an admirer of people with an amputation.

The attraction from wannabes into amputees can rest on the fact that a wannabe want to be like that person or to say it different –to have a stump around; by absent of having an own stump.

Earlier in this book I wrote that some wannabes want to reach their goal with the use of dry ice. Others try to do it with binding their blood circulation off. A very dangerous thing to do, but in that way body parts can indeed die off. So many wannabes told me that they did this with elastic, tape,… When the blood can't enter anymore in a limb then there is no oxygen. Without this a limb after a long time can die off. When I put years ago some inquiry's for wannabes online on my old Ampulove site, I received a very interesting and anonymous message:

- When I was a young child I found myself interested in amputated limbs. I was interested in above knee and above elbow amputations, finger and especially and toe stumps. I thought about my feet all the time and played with them whenever I could. I liked to bind my toes. I enjoyed the feeling and did not feel the pain when I tied them back to make the toes look amputated. When I masturbated this felt pleasurable until I ejaculated. It then became very painful and unbearable immediately afterward, and I had to release the binding quickly to relieve the pain. I found that this procedure made my climax much more intense; in fact it made it possible.

Through the years of adolescence I vowed to keep my horrible secret to myself. I was also interested in other limb amputations, especially above knee and above elbow. I was ashamed, embarrassed and felt guilty. A non sexual part of the body sexually stimulated me. I felt I should not be turned on by a deformed or missing part of another person. I kept this vow throughout my formative years until I was seventeen. Then I met my first wife, and I felt that I could not keep such a dark secret from a person with whom I was so deeply involved. I intended to share the rest of my life with her. Although my ex-wife had all her toes, she did have a strangely shaped foot, and that was enough to get my interest....

Here we come then again to the sexual aspect of wannabes. The ejaculation. More inquiries I did on wannabes gave me as answer that a big part of them can get turned on from the idea being an amputee, masturbate, ... And once they ejaculated, the wannabe feeling is gone for a little time. A very little time, and will return very soon afterwards.

If someone is interested in becoming an amputee, then we make also the link to feet. Surprisingly a lot of wannabes like feet. A European friend who was a wannabe and homosexual, became an amputee a few years. Strange is that he met his boyfriend on a website for people interested in the foot fetishism. When he became an amputee by wish, he never told afterwards to his boyfriend that his amputation was wanted. Even to me he told me one day before he became an amputee, that he was very afraid to lose his boyfriend if he would become a real amputee. This because they really liked each other feet so much.

Liking feet and at the same time being interested in stumps and amputations; It is indeed possible.

The same person that wrote about binding his toes off, is in the same way. He also wrote this:

- I have made wax casts of my feet and my wife her feet. From these I made latex models that look like her real foot. I have amputated some toes on the models. The stumps look life like. Even with all this at my disposal, I still want to go out and look for feet which are missing toes. I want to go to the beach and down to the French Quarter to look for feet. I look for women wearing lots of rings on their fingers or thumbs and ask if they would like a toe ring also. If they are interested, I give them a ring or two and ask for a photograph of the ring on their toe. Frequently I see women wearing toe rings. Many days I find no interesting feet or toe rings, and after looking for hours, I feel depressed that I wasted so much time and energy looking. Even if I found a good foot, what would I do with it? I can't take it home with me. I want to touch it and kiss it and feel it, but most women do not want that kind of attention. Besides, I am married and do not really want to fool around. Still, I am driven. If I do not go out looking, I feel I have lost an opportunity, one that I may never get back.

When wannabes try to become an amputee, and people from the medical world don't help them to become what they want to be, then some wannabes do sometimes strange things to themselves.

Years ago, a friend hanged for hours his leg in a freezer, so that after hours his calf was totally damaged. He didn't became an amputee afterwards, but he lost partial his calf. People, in the first place psychiatrics, named him someone that is into self-mutilation.

Self-mutilation is that someone hurt himself, just to be hurt. A wannabe don't want to hurt himself or isn't into pain. A wannabe just want to have his amputation and if he or she can't reach that point, then some of them will do things to themselves that can hurt, but not done with the idea that they would get pleasure out of pain. They only gone do this kind of things with the hope to get rid of the limb.

In the world of dreams, we need to know that a lot of wannabes have dreams during the night like every person have. A lot of them dreams about being an amputee, see themselves in a surgery room, how a surgeon removes their limb; And wakes up afterwards, back into reality and being disappointed, because it was only a dream.

Personally I know a few wannabes who had those dreams. The image of seeing themselves as an amputee. For me this is very clear that the need to become an amputee is so high and that it is indeed an obsession that only will stop, the day that the wannabe becomes an amputee. Proof of that is that the same group of wannabes with those dreams before, don't dream anymore about this when they have the amputation done.

Years ago, and also through the website Ampulove I put some articles online about *Herr Dr. Simon Schneider*. Dr. Schneider was a wannabe surgeon who had a private hospital, deep in the black woods of Germany. He was the only surgeon on the world that was able to do complete legalized surgery on wannabes, with even the help of the insurance of their country. I made the article in that way that they could see a painting from a surgeon, that they could ear a German voice who said that he was Dr. Schneider. One of the 'welcome messages' of the Dr. Schneider's page I still have from that time;

-Welcome to the real and only Dr. Schneider's wannabe clinic. A hospital with 30 rooms. All the patients came in with 4 Limbs, but never walk out with 4 limbs.. the most patients leave one limb in the hospital. Dr Schneider is the 'best' in stump creations. He did already over 3000 surgery's... off course only amputation's. Leg amputations are his most famous amputations. The surgery room is in the left tower on the picture. In this area, between the trees, wannabes can find rest, and peace together with their new changed body.

It isn't expensive to have a limb removed. A foot amputation starts already by 150 $, a complete leg is 1250 $. So almost every wannabe can afford such an amputation. Also promotions are running in doctor Schneider's hospital. Removing 2 limbs, a third one will be removed for free. Is this legal ? Yes, it is completely legal. even the German government offers you prosthetics and wheelchairs for free ! (doctor Schneider's wannabe hospital is located in Germany). So, contact doctor Schneider today, and get your leg of tomorrow !!!

This all was to find out how wannabes would react on this. I mentioned that the article was fake, but a lot of wannabes had not seen this, and were so excited with Dr. Schneider's story that they all started to contact him trough the site. It was surprising to see the messages from hundreds of wannabes, all emailing Dr. Schneider.

Some of them begged in the messages to see the Doctor soon as possible; Asked for more information, about prices, possibilities...

All sad to see how desperate they were, to read how high was the need in becoming an amputee.

I still have the messages from some of the wannabes who answered anonymous some inquiries years ago.

One of the questions I asked was:

- Tell us more about your wannabe feelings, try to explain us why you want this so hard.

Here are some answers I received that time on it:

- I want this, with the amputations I have reached already I am more happy already then i was before when I had still 4 limbs.
- Can't explain, just want to have the right knee disarticulated (and all the toes on the left foot). Already have right big toe gone (osteomyelitis), and will shortly have left big toe removed (same reason). Also lost tip of right little finger to frostbite (gangrene)and part of right ring fingertip to freezing.
- Like feeling of helplessness.
- I want it so bad because I fell my life is at a stand still and if I get i can move on in my life it is the only thing that i care about right now.
- It is nice for me to go on crutches and to make all things having one leg. It is so

> nice feeling as a big winning for athlete. Living with two legs is not interesting, grey and don't give satisfaction.
> - Hard to explain, I just know I find myself sexually aroused every time I think on leg amputations, specially above the knee.
> - Can't explain it, I just do!!!
> - The sooner I loose my leg, the better.
> - The sooner I got my amputation, the better. I already have muscular pains.
> - I was born with twisted legs. Wore braces until starting school, then bulky corrective shoes. Right leg is painful, I want to be free of it.

On the last two messages that people have send as answer on the inquiry I can say that there is indeed a group of wannabes that became a kind-of-wannabe, because they are in a lot of pain.

Some kind of disabilities and diseases can give a patient so much pain that he or she will start thinking about –Getting rid of the pain thanks to amputation. This group of people is not the kind of wannabe like described in this book. This group of people don't see an amputation as feeling complete, but want an amputation only to get rid of the pain, or to be in a better situation.

That surgeons are into helping people with an amputation with as solution a better live is not so much used in world parts as Europe or the United States.

In those parts of the world, surgeons don't think on performing an amputation if that could be a solution for having a better life.

In Brazil I know a young woman from the state of Rondonia who was born with a deformed foot. She walks on one crutch. Last year when I met her in person, she told me that a surgeon had told her that she would be better of with an amputation so that she could walk around afterwards in a normal way, with a prosthetic, without using crutches. A similar story I heard from a woman from the Philippines who had also a deformed and very short foot.

Also in her case a surgeon had told her that an amputation would be a good way to help her.

When you are a wannabe and you want an arm or leg off, mostly years of self study precede. Most apotemnophiles who go over to action are mostly well informed, most of them done already years of research about amputation and prosthesis, and know very well what they talk about.

When we asked the wannabes who cooperated at our internet-research the question whether they realized that phantom pains and stump pains might never stop we can say 97% of all the answers told us they were very well aware of that, and these pains wouldn't stop them in their wish for an amputation. Even if they knew beforehand that he or she would have terrible stump pains after the amputation, they'd still want it and go ahead with the plan.

When a wannabe says: 'I was totally happy the first day after the amputation', then this is almost unbelievable. We asked several ex-wannabes an open question, namely: 'Were you totally happy immediately after the amputation?'.

80 % answered no, only 20% yes. So one person out of five will be happy as of day one.

Remarkable is that the 20% exists of people who hardly had any pain after the amputation, the 80% comes from the group of people who suffered severe phantom pains, their answer to the question what could be the reason they were not perfectly happy immediately is clearly; 'because we didn't ask for this pain'.

When the pain disappeared, an ex-wannabe is perfectly happy and he or she will start an adapted life as amputee. As human you can't force someone to live with you, so our question is or -isn't it stupid to let wannabes suffer in their existence, isn't it better to help them, and give them the wanted amputation, isn't it their life anyway? They have to live with it. When one day a partner decides to live with a wannabe or ex-wannabe, then this is a positive point for the wannabe, and also a proof that wannabes can be with a non-wannabe partner. We know wannabes who got divorced, but also wannabes who found a perfect relation with a girl or boy who had at least tried to understand the deep wannabe feelings.

I know a few wannabes who live together with a non-apotemnophilia partner, maybe they don't accept the partner who is a wannabe, but a lot of them accept their wishes in becoming an amputee.

So many people have different opinions about wannabes. My opinion is pure based on my own experience I had with wannabes -the hundreds, if not thousand of them that contacted me during the last years. The different investigations and conversations I had with all of them. But it is always nice to look into other peoples opinions and thoughts about the world of apotemnophilia.

Marc D, wrote a few years ago a very interesting article about wannabeisme and an other view on it, with the title 'When less feels more';

-One of the most bizarre of psychological maladies is the overwhelming desire to have one's perfectly healthy arms or legs amputated. This once rare disorder, known by the psychiatric term apotemnophilia, appears to be on the rise. There have been various explanations of the meaning of apotemnophilia as well as explanations for its increasing incidence. But none of these explanations seems very cogent, as evidenced by the fact that clinicians have not been able to illuminate the dark feelings of those suffering from the disorder. - What, then, is the meaning of this psychological mystery? The key to deciphering a strange and apparently inexplicable compulsion is uncovering the normal desire of which the compulsion is a deviation, or perversion. As we shall see, the desire to have one's limbs amputated is a perversion of a fundamental human longing: the will to transcend the limits of egocentric existence through the act of self-denial.

The will to self-denial is not, in itself, perverse or crazy. On the contrary, it is the driving force behind all psychological maturation. To have the patience to be a good parent, for example, certainly requires self-sacrifice, an abdication of one's desire to control one's personal time, energy and resources. Indeed, the struggle to renounce one's egocentric mode of existence is the very meaning of an ethical or religious life.

Even self-denial in the extreme -- the self-sacrifice and self-mortification that approaches death's door, and sometimes, when necessary, enters the door -- is not, in itself, crazy. Such self-denial is the energy behind the noble asceticism of the saints, prophets, holy men of all religions, and of true philosophers. To understand self-sacrifice and asceticism, we must realize that it is not about losing; it is about gaining. The life of ego inevitably seems unreal. After all, we all know that our personal existence is something that will soon pass away.

Our knowledge that we shall die creates a sense of unreality that is experienced as a dread of death. It is also experienced as fault. Fault is the perception that our egotism has injured that which is truly real, namely the universe, the divine harmony, or something akin to that, depending upon one's beliefs. Only by sacrificing what is unreal, namely our finite and transient ego, can we gain what is truly real, in the sense of infinite and eternal. This longing to become unattached to what is unreal, so as to connect with true reality is the motive behind all sacrifice and self-denial -- normal or perverse.

Perverse Self-Negation - Many current psychological maladies are essentially a perverse expression of the will to the self-denial that belongs to ego-transcendence. Anorexia would be an example. The anorexic experiences any excess weight as egotism, and therefore as fault. Eating has always been connected, on a psychological level, with fault. After all, to eat we must take the lives of other creatures.

Most religions seek to sanctify and justify eating. Christians, for example, say grace before eating. The defense most often used to justify eating is that the plant or animal, by being eaten, will be converted into man's higher purposes. The anorexic apparently does not subscribe to this hierarchical metaphysics in which lower life exists to support higher life. Consequently, the anorexic feels unjustified not only in eating, but in existing at all. The anorexic must therefore display as little fat on her body as possible, for any fat seems indicative of gluttonous greed, self-assertion and egotism.[3]

Like anorexia, apotemnophilia is a perverse form of self-denial. It too stems from a sense of fault. Arms and legs, since they are our physical means for action, are psychologically connected to action in the world, to doing. But here is the problem: action that isn't justified, in the sense of being grounded in something absolute -- such as truth, good, the eternal, the Logos, God -- seems egotistical. Consequently, one cannot act without the risk that one will be plagued by a sense of guilt.

There are a multitude of psychological and physical disorders that are similarly connected with the inability to act. Existential neurosis -- the Hamlet-like impotence to decide and to accomplish, the incapacity "to take arms against a sea of troubles," due to one's inability to perceive meaning in the universe -- is in this class of psychological maladies. So is hysteria, the disorder -- rare today; more common in the 19th century -- in which a person's leg or arm became paralyzed, with no known physical basis. Catatonic psychosis, chronic fatigue syndrome, sexual frigidity and impotence -- when these disorders lack a significant physiological basis -- may have a related etiology.

It was one of psychoanalysis' seminal insights that neuroses -- particularly obsessive compulsions and fetishes -- are private religions. The person with apotemnophilia seeks self-denial, ego-abdication, self-sacrifice. But here the normal spiritual longing goes awry. Rather than seeking an inner psychological change, and thus becoming a "reborn" person, he short-circuits the energy of transformation by expressing -- on a primitive symbolic level -- what the spirit is requiring of him. He views the possession of all of his limbs as indicative of egotism. It means that he can act in the world freely, and self-centeredly, as an individual. Consequently, only by sacrificing a limb can he feel whole, because only then has he achieved oneness with the god he unconsciously worships.

The apotemnophiliac worships a Moloch, a god who demands life, or at least limb, from its followers. If the apotemnophiliac is to attain psychological and spiritual renewal, he must commit *theocide*. For only in killing Moloch, can a new god emerge in his soul -- a god of goodness, love and light.

We might add that just as a psychological malady like anorexia could only appear in time of abundant food, so it is that apotemnophilia could only appear in a time of peace and prosperity. The energy that would normally be used to oppose outward conditions is instead being turned, by the apotemnophiliac, inward against himself, self-destructively. Perhaps there is a certain wisdom in Freud's perception that therapy must free us of the self-destructive excesses of the superego by turning our life energies outward towards the world. Only in an age in which many people are awfully busy, but inwardly feel that nothing significant exists for them to do, could they conclude that their limbs have become superfluous. Consider, then, William James' cure for suicidal thoughts -- outward aggression against that which oppresses us. The apotemnophiliac's thoughts are not suicidal, but they are self-destructive, and so he should be encouraged to enter into battle against life's evils, rather than negating himself. When he does enter the battle, he will then feel that he needs all of his limbs.

Why the Energy of Transformation Short-circuits - Why does the energy of transformation short-circuit in the case of the apotemnophiliac and in many other psychological disorders? The key to neurotic behavior is that something is gained psychologically from an apparent loss. Despite the terrible suffering that the neurotic must usually endure on account of the emptiness of his life, he gets to stay himself. The neurotic person -- in love with his wretched self -- would rather commit suicide than let go of who he is.

In the middle ages, the church sold indulgences. An indulgence was a set fee accepted in advance by the church for future sins to be committed by the faithful. In other words, it was a bribe. Analogously, the apotemnophiliac attempts to bribe the inner voice of his spirit so that it will allow him to stay himself. The price he pays is a limb. Essentially stated, the apotemnophiliac pays with bodily sacrifice so that he does not have to transform inwardly. This transaction occurs, of course, on an unconscious psychological level.

It may shed some light on this matter to contrast true sacrifice with perverse sacrifice. A soldier who loses his leg in battle doesn't seek to lose his leg. He seeks, if he is idealistic, to bring about freedom, or democracy, or some other higher cause. The loss of his limb is an unfortunate result of the heroic life. It ends up being a sacrifice for the higher cause. Sometimes a deliberate sacrifice is required for a higher purpose, for example, when a person donates blood, bone marrow, or a bodily organ to someone else.

Perverse self-negation, on the other hand, is not a consequence of genuine sacrifice and heroism. Perverse self-negation grows out of a need to expiate fault. An example would be Oedipus blinding himself, when he realized, to his horror, that he had killed his father and married his mother. The film The Pawnbroker offers another example. The protagonist of that film, stabs himself in the hand out of an overwhelming sense of guilt and remorse for his cold heartless existence. The hand is symbolic, in the pawnbroker's case, of his grasping nature. Those who seek to have their hand or arm amputated have a similar sense of fault.

As we have said, what makes bodily sacrifice perverse is that it is really a surrogate for a true inner transformation. Oedipus blinded himself because what he had to see, the transforming insight, was too powerful to bear. His blinding was, symbolically, a regression to a level where nothing further could be seen. And as for the pawnbroker in the film by that title, Schopenhauer's words come to mind, "The only cure for mental suffering is physical pain." Of course, there is a much better cure for mental suffering: the illumination of one's desires, anxieties and conflicts. Self-illumination is the route to health for the anorexic, the apotemnophiliac, and all those who would seek to end their suffering.

The Problematic of Action in the World - The person with apotemnophilia, like those with related psychological disorders, is in flight from the problematic of action in the world. This problematic has two dimensions. The first concerns the will to be oneself, to exist. To exist in the world means that we must compete and contend with other people, and if not with other people, then certainly with other creatures and life forms. The second dimension of the problematic is not about fighting. It is about loving. It has to do with the need to give oneself to other people out of care, concern and love.

The first problematic, the will to be oneself, is embroiled with the perplexities of justification. Arjuna, hero of the Bagavad Gita, was in a terrible state of self-doubt and couldn't fight. After all, he was required to go to battle against those whom he cared about, his relatives and teachers. Cutting off his own arm, and so not fighting, would probably have been a welcome relief for him. But Arjuna's mentor, Lord Krishna, argued that refusing to fight would be cowardly, ignoble and a forsaking of his duty as a warrior. The only solution to Arjuna's suffering was to have his awareness ascend to a new level of answer to the problematic of action in the world.

The apotemnophiliac, the anorexic, the hysteric, the person riddled with the moral complexities and therefore rendered impotent to act -- all are troubled by this same problematic that Hamlet and Arjuna faced. But instead of transcending the problematic, those who are neurotic short-circuit their psychological and spiritual development through their self-negation.

The other problematic, we said, has to do with love. To truly love requires a shift of psychic energy beyond egocentricity, but the apotemnophiliac has found a way to abort his own self-development. He does this by turning himself into someone with a handicap. Consequently, he receives preferential treatment. After all, this is an age when people with handicaps are often viewed as heroes or as victims, or both; even if it is not understood, it feels right.

Contemporary egalitarianism -- which ennobles those who have suffered a loss because they are thought to be victims, and perhaps martyrs, of some social injustice -- may explain why such psychological disorders have been on the rise. All this makes the milieu just right to breed apotemnophiliac's. After all, the apotemnophiliac is not interested in loving. He is involved in a warped effort to be loved, and to him that means receiving. He receives economic benefits perhaps, maybe a handicapped parking sticker for his car, but more importantly, sympathy.

As a victim, the apotemnophiliac feels justified in receiving help from others, and he feels justified in not doing anything for anybody since his disability renders him unable to do anything. Without his limbs, he can only beg, or receive subsidies from the state. He says in effect, "I'd love to help, but just look at me." Most people who are handicapped are quite unlike the apotemnophiliac. They are more like the soldier who lost his arm in battle, in the film The Best Years of Our Lives. They are too proud to want other people's sympathy, and would be totally independent, if they could.

There is something rather comical in creating a situation in which one is rewarded for not serving. The novel The Good Soldier Svejk by Jaroslav Hasek (Viking Press, 1985) offers an example. The completely able-bodied protagonist, Svejk, pulls up to the draft board in a wheelchair with a fake cast on his leg, claiming to have a broken leg, and patriotically volunteers for service. Unaware that he is faking his injury, the draft board rejects him for service. The whole town treats Svejk like a hero for wanting to serve his country, on the front line of battle, despite his incapacity. Here, then, is a person who has refused to enter military service -- and, symbolically, a higher life of "service" to God and country -- but gets to be regarded as a hero. Of course, the apotemnophiliac goes one step further than Svejk -- he actually becomes a cripple. This is necessary, because Svejk is a charlatan, a person intent on deceiving other people. The apotemnophiliac, on the other hand, is involved in a self-deception.

To summarize, the apotemnophiliac is in flight from the problematic of action in the world. He can neither affirm his individual existence, nor can he transcend it through care and concern for other people. As is often the case with those who are mad, the outer man desperately seeks to mirror the inner man -- lacking justification for his existence, the apotemnophiliac wants the world to know that he is an "in-valid," and that he truly does not have "a leg to stand on."

When someone become an amputee, not as a wannabe, then a lot of them could end up in a depression with the idea that their life is over, that there is nothing interesting more to do. In this case, a lot of wannabes enter in the world from non-wannabes, who are amputees and can be in a lot ways a positive help for the depressed amputee. Well known is that the wannabe who became an amputee only can talk positive about live as an amputee.

The depressed amputee who never wanted to become an amputee can see this as a positive push in the back. The wannabe-amputee that tells to the non-wannabe amputee that live is so beautiful, even with having a stump, that lives goes on, that they still can do everything...

A lot of amputees appreciate the fact that someone push them in the right direction of having a good live as an amputee. In the most cases the amputee will never be informed that someone who wanted his amputation(s) helped him to become back a positive oriented person.

An other view on the wannabe phenomena was founded years ago in another text I found years ago:

- There are cases of self-demand amputation (Apotemnophilia). These are often related to the erotization of the stump and a sense of over-achievement despite the resulting handicap. Such an obsession tends to represent a kind of "fixed idea" rather than a paranoid delusion. It may be conceptually related to, though not identical with, transsexualism, bisexuality, Munchausen syndrome and masochism. As with most pharaphilias it is far more common in men. There is limited knowledge on the relationship between sexual attraction and amputated limbs, but there is clearly a syndrome of erotic obsession or fetishism for amputated limbs or digits. Many see the amputation as an erotic fantasy which becomes a potent, obsessive desire. Photographs of seminude or fully dressed amputees often serve as a visual masturbatory aid. Gender identity concerns can lead to these symptoms, although self amputation of the penis is sometimes substituted with the leg, to avoid going the whole way. The resulting stump suffices as a kind of surgical masochism. Contact with other apotemnophiliac's can act as a half way house. However, the obsession is difficult to overcome. Paraphiliacs generally wish to transcend an aversion or the forbidden.

A pharaphilia is a condition occurring in men and women of being compulsively responsive to and obligatively dependant upon an unusual and personally or socially unacceptable stimulus, perceived or in the imagery and ideation of fantasy, for optimal initiation and maintenance of erotosexual arousal and the facilitation or attainment of an orgasm. In Acrotomophilia (amputee fetish), the specificity is that the partner must be an amputee. An acrotomophile is erotically excited by the stump or stumps of the amputee partner and is dependent on them for erotosexual arousal and the facilitation and attainment of orgasm. People may become attracted to amputees for many reasons. These include fetishes, vicarious sadism, low self-esteem, legitimate excuse for poor performance, respect for an amputee's adjustment to an handicap, need to rescue and care for the amputee or even fear that they could face the same affliction. The amputation is an obsession that applies usually to the self. It can occur that the acrotomophile seeks to be amputated himself. In the absence of self-demand surgery, an amputation may be self-induced, as an apparent accident for example that is completed professionally in a hospital. Such acrotomophiles are sometimes referred to as devotees. Most acrotomophiles seem attracted to the stump, although sometimes severe scarring is the turn on.

How the amputee presents him or herself is important. Broken off or crippled limbs seem to be less exciting and most of the amputees try to show off their perceived better points, by strategically choosing and arranging their clothing. An additional feature of amputation/amputee pharaphilias is that they have strong social as well as sexual components. Acrotomophiles obsessed with the amputation of an intact person typically find that his/her pharaphilias obsession is not reciprocated. It is questionable whether any of the other pharaphilias can come close in terms of being twenty four hour fantasies. Whilst it

103

remains a fantasy, twenty four hour or not, it is potentially transient, but once mutilation becomes reality it is generally permanent and irrevocable and the next stage can be death.

Before the first surgeon who ever did an amputation on wannabes, a lot needed to happen. Like I mentioned before in the book, a few years ago a surgeon in Scotland performed amputations on wannabes. Before he started with that he came in contact with a man who was a wannabe and found Dr. Smith to do the amputation on him. A single above the knee amputation in his case.

The man who became an amputee thanks to the help of Dr. Smith had for many years the internet site *Overground*, with very interested articles on it. We can say that it was one of the first sites with wannabe oriented information.

So many wannabes published during the years their stories in magazines, internet and websites. But I still remember very well the more then amazing story from Peter, who was for long a friend and even visited me years ago a few times. Here is the wannabe story of Peter and what happened with him:

... It all started when I was seven. In my class was a girl named Helen. One day the teacher entered the class and announced: 'Helen is in hospital, she had a leg amputated'. At that moment I got a strange inner feeling, as a seven year old I thought about how she would look like, on one leg, how she was lying there in the hospital, and I realized I was jealous. At night when I came home and was lying in my bed I couldn't think about anything else then Helen, I also wanted to loose a leg, and what came mostly in my mind was the question: 'Why isn't it me who is lying there in the hospital with one leg off?'.

That is how it all started, after a while, as I grew to adultery, I started to do more research towards amputation. I discovered more about different amputation-techniques, but never something about the fact why I wanted a leg off. I got married in 1989, and never told my wife about the fact I had wannabe-feelings. I was enormously shamed over the fact I wanted off my left leg above my knee, and at that moment I wanted nobody to know about having this dream.

When I was 22 years old I found in a book that sometimes limbs were tied. I took an elastic ligature, and cut of my arteries for hours, in a way that I was rolling around out of pain. Meanwhile I had been since I was 12 several times with my leg in plaster. In total about 20 times. I always found a method, so they had to plaster it. I did this because I liked the feeling not being able to move my leg enormously. When I was married a few years, and I was about 25 years old I decided to apply another method. I also found in a medical book that freezing could lead to amputation. In the basement at home stood a large freezer, I decided to hold my calf against the freezer-wall for a few hours. After a while my calf got really red and hard from the freezing cold. After two days it was black and dead. I didn't told my wife what I had done, I told everybody I didn't knew how it happened, it was suddenly there.

At arrival in the hospital they asked whether I got in contact with dangerous products, but I never told the truth. After two weeks of hospital my leg was still on my body, only my calf was as good as amputated, and a few slices of new skin were attached. The most horrible moment was when I woke up and found out that all my tries I had went through were for nothing, and that stupid leg was still on my body. Indirectly I tried to talk about it with my wife, I often asked her about amputation, whether she would stay with me if I ever would lose a leg. When I was 26, I got internet. I found medical texts about wannabeisme, I discovered I wasn't the only person in the world who wanted to lose an arm or a leg. This feeling gave me new courage to continue my search for perfection. One day I decided to write anonymously to two well known orthopedic surgeons, with a short explanation and a request to meet at a certain place with me. Off course neither of them showed up. I thought that now I discovered on the internet, I wasn't the only wannabe, I could tell it to my wife, I printed out the medical texts I found, and gave them to her as a sign. I told her I was like that, that I wanted my leg off for years. We didn't talk much about it, it was hard to believe for her, her husband was an apotemnophile. After several conversation it came to the fact she didn't want me to amputate a leg, but I could have a small toe amputation. I informed my doctor who came to our home almost for fourteen years, and of course he was more then surprised when I told him with what feelings I was walking around already for years. At first it looked all very unbelievable, as well to him as to my wife, no-one would talk to me about it, and that was just what I wanted, someone who wanted to listen to me. That time I had a good friend, I told him that I was an apotemnophile, and without thinking he told me that it had something to do with amputation, he studied and knew the meaning of the word apotemno, and started to realize I really wanted a leg off. He answered me: 'I should have known this already'; During all our friendship he saw me often enough make drawings about amputees, or in school when we were in the same class, saw me sitting in the wheelchair of the nursing class and that I really liked to roll around in it through the corridors. But I didn't get much help from him, he understood, but when I tried to talk about it in later meetings it was often that he asked not to talk about those wannabe-things.

Good, I had the approval from my wife to amputate a toe, a small toe, by freezing it, so they had to amputate my toe, it was better then nothing, so I decided to start looking on the internet for more information about toe amputations. On that occasion I encountered a German wannabe, who amputated himself ten toes (and had kept his amputated toes in the freezer of his home during all the years). He told me how easy it was to cut off a toe yourself without pain or danger. One night I decided, I couldn't wait any longer, for years I wanted an amputation, the moment to get rid off a small toe had arrived. While my wife was sleeping I went down to the bathroom carefully. Once there I assembled all the materials needed to get rid of that toe. Ice from the freezer would anaesthetize my toe so I wouldn't feel any pain. And yes, the more ice I put on my little left-toe, the more it got sleepy. After about ten minutes the toe was asleep and I realized it would never ever ..wake up.. again. This alone was already a wonderful feeling for me. I took a cutter, and cut through my toe, at the place I wanted it off, just above the fold. It was like cutting through butter.

A few moments later I noticed that the toe turned white, and the only connection with my foot was the bone, and it started to feel cold. Now I realized that when I cut through the bones I really would have amputated a toe. A few moments later the bone was cut through and I had my little toe in my hand. I tied up the wound, just as the wannabe on the internet told me, and cleaned up the small amount of blood that was left. I made a note for my wife so she knew why I wouldn't be at home for a while, and left to get rid of the toe. I threw him in a river not far away from home. When the sun came up I returned to home, and told my wife what I did that night. Even though she told me before I could have a toe amputated, she wasn't really happy now. She never thought I would be strong enough to amputate my own toe.

The small toe-amputation healed soon. Afterwards I only regretted that I didn't feel that my toe was off. I really wanted to feel the good amputation feeling. I wanted to feel my stump. I decided to amputate another toe, this time without informing my wife. I thought that when this time I would cut of my big toe that I would be able to enjoy the feeling of amputation. It was during the day, my wife wasn't at home, everything was ready again in the bathroom, and I decided to cut off my left big toe. This went even easier then the small toe I did before. Later I got in the car, went to a friend, who studied nursery and I got to know better and who understood me and tried to listen to my wannabe-feelings. I didn't dare to tell it to my wife myself, so I asked her to come to me, so I could tell her what I had removed this time. After the necessary explanations I went to lay down on my bed so the bleeding would stop finally, but it didn't stop and soon I realized I had to go to the hospital. On arrival there, my family doctor already called to give the necessary explanations, but despite that it wasn't nice knowing the whole department and medical staff knew I had cut of my own toe.

The other day I was so tired of it that I got in my pajama's and in my car and drove home. I enjoyed my big-toe amputation enormously, it was a wonderful feeling, but I realized I still wanted my leg off. I tried to get my friends to understand it, but later I heard from others they were laughing at me behind my back. My wife also got tired hearing me talk about amputation all day long, that tired we decided to get divorced. This was hard for me. I really loved her, I trusted her, and I shared my deepest secret with her hoping we could talk about it as man and wife. When you get married you promise to be there for each other in good and in bad times, but that wasn't the case now.

When I lived alone in my rented apartment I got more and more in contact with the other friend who listened to me, not that she approved what I wanted; having my leg off, but at least she paid attention to me. We fell in love and kept coming closer to each other more and more. One day I went to the bathroom and amputated two other toes, the day after that I amputated also the middle toe. My girlfriend who actually had to go to school stayed with me during this period with infernal pains that I had to go through that week.

Afterwards, with a foot without toes the feeling was *terrible* nice, I felt great, but yet I couldn't get rid of the dream to live one-legged. Now I was getting divorced, and lived alone, I decided to carry on with my plan, no-one could stop me, not even the idea my son would have a one-legged father.

I tried the freezer again and froze my leg again, and decided to inject a self-made infection with urine, death flies and other garbage into my leg. I only got sick, and my doctor only would give me painkillers, and wouldn't do anything else. After being at home for a month, with a leg that stank because of died tissue, and heavy pains, my doctor phoned me and said he wouldn't subscribe me other painkillers. I started yelling at him and an hour later seven police agents came to get me at home, at first I thought to bring me to an hospital to operate the calf again, but no, they brought me to a madhouse.

There they told me I was collocated, I didn't had any rights as human being, and I was some kind of *state property* without any rights. The first night I could stay in a open place, all other rooms were taken. I sat between crazy people who thought about nothing else than suicide, drugs and alcohol. Behind iron bars I realized the day after my parents-in-law came to talk to the nurses. I was locked up between four walls, no fresh air, only fools beside me who wanted to be death. I slept in a room with an old fool who was one-hundred percent *nuts*; He had millions of dollars hidden under his pillow, and he had to guard it all night long; That was what he told me, and how he kept me awake. The day after, my lawyer came to visit me, I was so hard stoned from the medication they had offered me, that I didn't recognize him, and asked ..who are you?..

He told me it wouldn't be easy to get me free, and if he wouldn't succeed, I needed to stay there at least a whole month. On the forms made by the psychiatrists stood all lies, according to them I was someone who thought about nothing else but suicide and self-mutilation. It was terrible hearing this, knowing I didn't want to die, I only wanted a leg off. Meanwhile, they decided to get me to a local hospital, for an operation on my leg. The surgeon there was very unfriendly, he even said in a sarcastic way: 'Wouldn't you rather have your leg amputated?', I answered him: 'Yes, of course', but he went angry as hell. I told him I smoked, but he answered that during my stay there I wasn't allowed to leave the room. I knew I couldn't deal with such a situation. My case came up in court a few days later, and when I would stay there it would only get postponed, and then I should stay even longer in that madhouse.

I decided to escape, pretended to go to the toilet, but run through the hallway like crazy, left the hospital and lifted without anything to a town about eighty kilometers from the hospital away. There I decided to go to my girlfriend. I called first my ex-wife. She begged me to return to the hospital, because the cops got my son from school and they thought that my ex-wife knew where I was. I called the family-doctor, he -who had done this all to me, and I declared that they could come and pick me up, on the condition that my wife would get our son back, I didn't have to be locked up in a cell, and my trial came up a few days later.

Later, on the trial, my wife was sitting next to me, the judge in front of me, beside him my doctor and beside him the crazy psychiatrist. My lawyer talked about the fact that transsexuals also want an amputation, and that amputating a penis was something like wanting a leg off. He also said that they would let me free if I should visit voluntarily a psychiatrist. The most important point though was that the date of my collocation was a day later than when the police agents came to get me. This made it that I had been locked up one day too much.

That night I got a phone from the psychiatrist. He told me that I had won the case, and could leave his hospital immediately. I couldn't believe my ears, the worst week in my life was finally over. At home I broke out in tears. I was a free man again, a free man that was damaged in a terribly way by his doctor who I trusted and knew for fourteen years, a man who promised me to never change my wannabe feelings. And still I decided to keep him as doctor. Some time later I went to visit the psychiatrist. This psychiatrist told me he thought that *internet-thing* was some kind of *sect*, and internet was the cause of me -continuing my plans to get rid of my leg. I always kept saying that I didn't want to loose my leg anymore; Just to get rid of him.

Meanwhile I got to know on the internet a doctor who was a wannabe himself, and explained to me how easy it is to paralyze a leg with only local anesthesia. The psychiatrist told me in the meantime I wasn't a wannabe anymore, and didn't had to continue going to the sessions. One month later I was in a neighboring country, visiting that wannabe-doctor who put my leg asleep with one simple injection. Afterwards I drove back home, but the wish for an amputation was that big I cut in the stump of my big toe while my leg was sleeping, so I couldn't get home. I realized this was the perfect chance to become a real leg-amputee, I realized that if I just went home I wouldn't be able to get this chance a second time. I returned at night, bleeding, back to that doctor's house, and as my leg was still under anesthesia, the doctor closed the stump again. I begged him to help me, to give me a plan, so I could finally loose my leg. I thought up a fake accident, found a nice place where I could do my amputation. The evening the amputation was planned I returned to his house, and again my left leg was put asleep. When leaving the house, I broke my ankle but just continued walking. I realized it was the last time I felt my foot. On the place of the accident I chopped off my own foot with a hammer, three minutes later my foot was off. I took it in my hands, kissed my amputated foot and said: 'You'll never come back on my body.!'. I hide the tools I used to get rid of my foot in a horrible way.

A few moments later in hospital, I told another story and made everyone believe I had an accident. Fortunately everyone believed me, and the next day I woke up with my left leg amputated under the knee. Finally I reached my goal, or at least for a large part. My dream had always been to loose a leg above the knee, but before I went abroad to set up this fake accident, my family doctor told me he never wanted to see me with an upper-leg amputation, this because walking would become very difficult.

So, afraid to be locked up again, I didn't go further then an amputation below the knee. Soon I got a prosthetic, and learned to walk again. It felt great to go through life on one leg, but the knee really was still too much. Off course I couldn't get into another accident, it would be too obvious. A new plan was ready... I decided to go to several surgeons in my own country and told them that I had terrible stump-pains and pains in my left knee. The third surgeon believed me. And only two days later I was in the operating room and heard the *sweet noise* of the saw that would amputate my left leg completely and for good.

After the amputation I immediately went home. I had what I wanted for years, a real left upper leg amputation, and decided to recover at home. A new life started for me, a life as amputee, a man on one leg. Looking at it later, I had to go through hell, and did many dangerous things, before I got what I wanted. The *sorry*-thing about it, is that I didn't loose a leg when I was seven years old, when I had this wannabe feeling for the first time. This feeling of amputation that I wanted to have many years before, -It feels just the way I always imagined, and I enjoy it daily. It's very nice when I try to move my leg that is no longer there, and luckily won't come back. I feel sorry for the others, other wannabes who want what I always wanted, most of them will never get what I got, and most of them aren't that strong to do what I did. Most of all I'm furious at the medical world for their lack of comprehension, and because it is their fault I had to do all this to myself, when there was acknowledgement or no taboo, I shouldn't have done those things to myself which almost cost me my life. I don't regret what I did. I enjoy every day of my new life, and I hope other wannabes will also get what they want.

After the story of Peter, who I met a few times when I still lived in Europe, I met more wannabes and heard more stories, but about myself, my own experience, to end up this first part of the book, then I need to go back to about tree years ago. I was sitting in front of my computer and was talking with some wannabes. I discovered that a few of my wannabe friends lived very close to my house here in Florida. Why so many wannabes live here in the area wasn't clear for me. Maybe because I don't live that far from Miami, a big city. When you live close to a big city, then there are off course more wannabes.

Near that I had a good friend from a state more to the North; He decided to come to visit me. Already years I knew him from the internet, but we never had met before. He was a wannabe, became a multiple amputee and he helped a lot of worldwide friends from him with small amputations in the way of doing *home surgery*. When he arrived at my place, he told me so many stories about all the wannabes he knows. He told me about a guy from Mexico that even went to him to have tree fingers removed. Later it went that night, more people came over to my place. All people we both knew from the internet.

People?; yes, but all wannabes. It came to a point that later that night we were eating a big pizza with four people. All amputees, and the most of them? –*Amputees on request*... Wannabes. One of them had a toe off. An other had lost a part of his foot off, then there was my multiple amputee friend who helped wannabes, and then there was me. Never before I had believed if someone had told me that one day I would be in my house with 3 wannabes with me. Four wannabes if you include myself as a wannabe. And about myself as a wannabe; In the beginning of this book I wrote about it and again I would go a little bit deeper in on it. When I had my accident, and later needed to follow more amputations, I indeed didn't care about it. Basically I was already used on a life as an amputee, and I could not see the difference between having one or more stumps. I was and staid an amputee anyway. The fact that I came in contact with the devotee and wannabe community afterwards, made me feeling good in the way that those, -in the first place devotees –later the wannabes, made me good feeling. They helped me to make myself strong, to give me the feeling that even as an amputee I could do everything I wanted to do. From there also maybe the appreciation I have for devotees and wannabes. For sure I am not the only open minded amputee, open minded for devotees and wannabes. Wannabes are just people like everyone.

They don't want to harm anyone, even not themselves. They just want to be happy in their life, in the way they want. So many people can say now: 'But they are not normal, they are *nuts*!'. Let me interrupt then with the question: 'What and who is normal on this planet?' and –Who are we to decide about what and who is normal...

During all those years I got involved in the wannabe and devote world as an amputee, it was maybe normal that a big part of some people started to think that I am a wannabe. Even worldwide came some freaky articles out from me.

Two years ago an English media group contacted me, asked me for an interview about me and my contacts with the wannabe and devotee world. Good believed as I was I gave them a lot of information and even pictures. But later when they start to sell worldwide their article about me, it was pure as –Alex the wannabe oriented. I was upset in the first place, contacted a lawyer specialized in media cases, but already soon I found out that it was very difficult to start something against them. So, I didn't care anymore about whatever they wrote about me. Last year an Australian radio station asked me for an interview; Also about me as a wannabe. I told them that a disturbed

119

woman who was only into article-hunting sold my story worldwide, a fake story. The Australian radio station send me later that they a scan of an article from me that came out over there. I needed to laugh, because somewhere in the article was mentioned: 'Good that we took the interview from Alex now, imagine we would had interviewed him a year later, then probably nothing from him had been left over, and interviewing only an eye or a finger, while all the rest will be amputated, would be very difficult...' I really didn't care what was written in that sensation magazine. And I refused any radio interview, because that Australian woman wanted to keep it only wannabe oriented; This in the way that I was the big wannabe.

A good year ago an other article from me came out in a big magazine in Europe. Strange of all, it was written by a friend who was a journalist and had contacted me before. He wanted to write about me and my world travels. But when I received a copy of the article; four pages long; I needed to discover that 2% was about me and my travels and about 98% about me and my amputations. I told him that I had from my amputations medical proofs and that if he didn't believed me, he could always contacted my second wife who was with me when I had my accident. The funny part was that I heard afterwards that justice had decided that all those magazines needed to be taken out of all the stores that week. Indeed for my article, but not for me. That wasn't the reason. The reason was because they had published the article not only with my pictures they had token without my permission, but also from a female amputee. They had used her picture without permission and she had started a case against the magazine.

The most magazines and television channels are just hunting for some articles. If they have something to write about; Something that chocks the whole world, then they are already happy. The wannabe world is for sure a thing that media is interested in. A lot of my friends were hurt and used thanks to those media-horny hunters.

When years ago, I helped the BBC, for their documentary Horizon, it was the same thing. They didn't used me, but they for sure used the wannabe woman who was in it, that became an amputee afterwards, without their help.

That the world of wannabes is a very closed world was again very clear during the last days. Through a few internet communities and Yahoo groups on the internet I tried to get in contact with wannabes. I published a message with the request to wannabes to get in contact with me. To give me more information about their experiences, stories, etc... Unfortunately, no any email came in, even not with mentioning that their story would be kept private.

* * *

Devotees and Acrotomophilia

Near wannabes, who want to become an amputee or disabled, we have the other group, called Acrotomophiles or devotee. Devoted to people with a disability. The devotee group on this world is with much more then the wannabes are and some old statistics say that one on five hundred people is an admirer.

As being a triple amputee I need to tell you that for sure there are much more devotees then only one on five hundred. From own experience I can say that I was in contact with a lot of people interested in me and my disability.

Years ago, after being a left above the knee amputee I met Angelique. A woman from Austria who was obsessed by the idea to meet me, to be able to be together with an amputee. She took the train and came to visit me in Belgium, were I lived at that time. In the first place it was nice to have a woman around who didn't mind that I had only one leg –that time. Even when we went out in public, she always put her hand on my leg stump and didn't mind that people were starring. Angelique was the first female devotee I met after I became an amputee.

On all those years as an amputee, thinking on the partners I had, I need to confess that probably 95% of them are devotees. Years ago, when people talked about the admirers of amputees and people with a disability, then a lot of disabled people thought that devotees are sexual obsessed by amputation and disability. That they are only interested in the sexual aspect of it. Let me say that a devotee is for sure not obsessed by sex. Not only the erotic part is interested for a devotee. Years ago I picked up the word '*Stumplover*', but then the question is –What is a stumplover, and how far does it go? ...

A stumplover, let us use the word devotee; -Someone who is attracted to someone with a disability. The biggest part is in the first place attracted to amputations (stumps), after that to people who are in a wheelchair, use braces, more...

That having sex with a *cripple* is the only thing that a devotee wants isn't truth. A devotee will think in the first place that he or she find the disability attractive. They can like the disability that a person has, but they see that person as a whole person. A stump, a wheelchair is just something extra that turns them on.

Let us take as example the attractions in a way we all know. So many people are attracted to blond hair, blue eyes. While other are interested in interracial, having a black or white partner. Attractions can be so different. There basically need to be an attraction to something to be with someone.

An attraction can be something simple. Having the same kind of interests is an example of intern attraction to someone. Near the intern attraction we have then the extern attraction. Finding someone attractive. But to find someone attractive their need to be something to find that person attractive. In the case of admirers, well... they find the disability very attractive. Finding a disability attractive sounded years ago as something that wasn't acceptable. Now those days it seems like the world have accepted much more the fact that a lot of people like disability. A good example about how open minded the group of devotees and people with a disability are? Let us take facebook.

Those days, everyone almost have a facebook. A perfect way to get in contact with old school friends, but also a very good way to get in contact with amputees and admirers. On my own facebook, for sure the most friends on it are devotees. So many times I receive messages that they find me attractive, and without asking why they find me good looking; -I know why.

Now, coming to the gender of admirers, we can say that there are at least so much female devotees as male acrotomophiles. In my own situation, looking back at all the years I am an amputee, yes... my girlfriends were mostly devotees. For me it is very clear that someone who is together with someone with a disability, need to be at least a little bit interested in it. Why should someone be with a person with a disability, knowing there are so much good looking people around who are not wheelchair bounded, who still have all their limbs. Why should someone decide to be with someone with a disability, knowing it is much more difficult to go out with someone then a disability then someone that just can run around on the beach. Looking at myself, I don't run around on the beach, I still can do everything, but the most things in a different way. Well, my opinion about this is that everyone I was with during the last years was a devotee.

Devoteisme can go far. Some people really need to be with a disabled person, while an other group just likes disabilities, but don't have a real need for it. The Angelique I told about was a good example of a woman that needed to be with someone who have a disability. In her case an amputee. When we are talking about the group of female admirers, let me say that the most of them don't say from the first day that they are attracted to it. Same as with the female wannabes, what I mentioned before. Even once, a woman who was for sure a devotee, was upset because I told her that she was an admirer. She told me that she wasn't. But a few weeks later she contacted me again and told me: 'Yes I am, but I never came out for it before'.

After my second ex-wife, I met a girl from the Caribbean. Her name was Nathalie and lived in Brussels, Belgium. She told me that she was very attracted to male amputees, that during her whole live, she was dreaming about being together with an amputee. Well, we went together and lived for a while together. During the sexual activities, it was much more interesting. –my opinion.

Not only I was interesting for her, but also my stumps were. While I realized that my stumps would discuss a lot of woman.

After Nathalie, I was in contact with a woman from Norway. She was together before with a man who was wheelchair bounded. Not an amputee. She told me that she wanted to visit me. In her case, flying from Europe, to here in the USA.

She informed me, that for her, being with an amputee, was more important then being with a man in the wheelchair, who wasn't an amputee.

But, I never met her.

Once, I had a girlfriend who found it such a warm up, when I hold my arm stump on her pussy. She became so hot from it, that at those moments, she was more horny, then that it would had been a penis.

Sex in the wheelchair is another thing that some of the devotee woman requested me to do.

Once there was Ann, a very nice Belgian woman, who had four children. Her marriage was over and we met through the internet. She wasn't a devotee, but for sure she had no any problems with the fact that I had (at that time) only one leg. Some woman are just interested in being with such a partner because it is different then the usual thing.

When I went out one day a few years ago to Hooters; -a sport bar, known for all the woman that work there, because they are young and beautiful;

I met a girl that worked there, Krista was her name. Krista told me that her sister died years before, and was in a wheelchair. Krista gave me her phone number and a few later I was dating a very beautiful Hooters girl.

Was she a devotee? Looking at the fact how beautiful she was; -She could have every man on two legs. The fact she was very interested in wheelchairs, ... yes.

Afterwards I met my American wife. Before I married her, she told me already that she was a devotee and very interested in man with a disability.

I also met her through a website that only was oriented in disability dating. But even the stumps didn't kept us together. For me a proof that for a relation there is more necessary then only stumps and disability.

Once a female admirer wrote on the old Ampulove site:

"I would spend hours, just thinking of caressing a man's stump: the way it would feel, smell and taste."

So many times I heard in all those years that a relation was over, because one of the partners became disabled. Husband and wife together; Perfectly happy with each other, till... one day, one of both becomes disabled. My own accident was a good example for this. After it, my wife could not go on living with me, so we got divorced. For that, we maybe need to say, that it is a good thing that devotees exist. A group of people that like people with a disability and have no any problem to walk over the street with them.

During many years a lot of internet sites offering amputee oriented materials. Basically pictures and movies. For a certain amount, people can buy pictures and films from amputee models. One of the first companies that ever started with that was *Ampix*, but is already a few years offline. Near Ampulove, the site I had for many years, there is Ascot World, a site owned by an American amputee woman. She even organize amputee – devotee meetings.

About those meetings, I can say that there are a few companies, mostly based in Russia, organize meetings with amputated Russian woman. A male admirer can find in that way a date or relation with the woman of his dreams; - An amputee. The prices to meet such a woman are mostly very high.

Those days, there are so many sites who offers materials from amputees. I did a lookup into this. Sites as Disabledplanet.com offers about 1200 amputee and disability movies, over 65.000 amputee and disability pictures; all online and ready for download; - stories, devotee information and so much more. Afterwards I found out that this site is owned by a female devotee who's name is Melissa.

Near Disabledplanet, the most devotee sites are based in Germany.

Even after being in contact with a Brazilian female amputee model, the answer on my question: 'For who you work?' was: 'My manager is from Germany'.

That an amputee model is unique and not that easy to find is clear when you look at the prices picture sets, movies and memberships are sold.

A movie $100, 25 pictures for $49.95, are normal those days.

Since there is YouTube, Flickr and other ways to share pictures on the internet. A way to download for free amputee pictures and movies, less people get interested in those sites like it was years ago.

I remember that with the Ampulove site I had years ago, there was once a day with over 12.000 visitors. Indeed very much, if you think about that this amount, are only persons interested in amputees.

After a look up on sites like webdetail.org, where people can see how many visitors a site has, then I came to the conclusion that this is different then it was before.

Near the fact there are so many amputee and disability pictures and movie sites; There are also a lot of dating sites. Dating4disabled is an example. It is a way how people with a disability can come in contact with a non disabled person.

And coming to the subject sex. Even a few years ago there was a German one legged woman who did escort for man interested in stumps.

Years ago I was in contact with Rose.

Rose was an above the knee woman from Australia.

She didn't know that there are people interested in amputations. She told me that she had sex with man for money. A few days after I told Rose, that there are devotees, I found already advertises about her interest to meet devotees.

Some devotees really travels over the world to meet an amputee. It is not that weird if a person from Europe or USA would fly to Austria to meet an amputee.

Years ago there was the cartoon Amputee Love. Probably the first Cartoon that ever came out when it comes to the subject amputation and the sexual fantasy around it. Those days, so many admirers worldwide, are making graphics, cartoons and even oil paintings, that shows the amputee in an attractive way.
Near the escort and paintings we have the professional film industry. So many movies with an amputee in it, but also pornographic movies. Years ago Jane Silver was an actress in a pornographic movie that came out in the USA. In the movie you can see how a man licks her below the knee stump and how she hold her stump on his ass.
A year ago a Brazilian movie producer from Sao Paulo found me through the internet and asked me to play in her movie. She informed me that she was a professional moviemaker, connected to an American company specialized in porn movies. She made all kind of movies in this style. She had find out that there was a big interest for the pornographic

market. She offered me $3000 to play in a movie, to have sex with six different pornographic actresses. But I wasn't interested to show my naked stumps all over the world.

Not only from her I received weird requests. But a few other people tried to get me involved in this kind of business. At the time I still lived in Belgium, I received once a letter from someone that told me that he was a film producer (See letter on next page). First I was very exited, but more I got in contact with him, more I discovered it was a devotee only

To: Alex Menasert
Belgium

From: Larry
Berkeley, CA 94703

Howdy!
    I wish to approach as a friend, in honor of my great-grandmother who was a **quad amputee** from diabetes, a nice slender blonde Dutch girl Maiden-named Wilhelmina Schrően (pronounced Shren). She never hid from people, and even acted as the <u>**Nude Maypole**</u> at the local <u>Spring Equinox Dance of the Ribbons</u> around 1910! This custom <u>must</u> be revived, and why not in a **Musical Movie**? It would <u>really sell a movie</u> (**big $**), being <u>based upon actual events</u>! (With all ribbons fastened to the high collar of the **nude "maypole" QUAD fertility goddess**, the ribbon-holding female dancers first danced around in one direction 3 full circles, then back the other way 3 circles, then <u>resembling the Virginia Reel, wove</u> a multicolored ribbon dress-skirt about the "maypole" with half the dancers going clockwise, and half counter-clockwise, <u>weaving in and out of each other's path</u>, to a jaunty tune!) The public absolutely loved it! After Minnie got too old, nobody else had the guts, or the amputations, to maintain the custom in a small town, which even now still has a gleam in it's eye. (I envision an amputee sub-community, one cute **quad girl** volunteers to "maypole", and revive the old custom. Her lover is of mixed feelings, despite "SAK Iraq", the town is divided, other towns copy it. **<u>Many subplots with "something each" for most separate demographics</u>**, and on and on.) As a musician, I can hold up that end of things, if you want.

    I believe you are the best person to be at least a rallying point (if not a hood ornament) for this movie to be made. For one thing it would solve all the financial problems, currently being faced. Some of my show-biz friends have been wanting a project with me like this, for a long time. On that basis alone, success is assured. No BS. I am a singer-songwriter-guitarist with a hand in much successful music, and Bob Seger is my cousin.

    And on to current **business**. A few days ago I tried to send Stumpophilia/Ampulove $4.50 by credit card and join up. I am under the impression Stumpophilia and Ampulove are under the same organizational umbrella, but if I am wrong, it won't be the first time, so please educate me. As my attempts to leave a message on your site, keyed to my e-mail, to offer to receive merchandise to rectify the situation have failed, I now make the offer by snail mail. **Just send me a copy of Dream Stumps, and credit my folder with as much membership as you believe I deserve**. I will call it even, and look forward to working with you and associates on this **musical** movie. A genuine place in History.

    Even President Bush is now supporting amputees on national television! Also see the enclosed clipping. The public is becoming curious, and **now is the time** to move forward with streamlined business practices and <u>media presentations</u>. The time of murmuring and veiled shell-game style presentations is suddenly over. This new tide if taken at the flood will float a lot of boats! Big movie **financiers** await a working script **outline**, the actors are already known to us, and ready to "play themselves", so to speak, no acting ability needed. I suggest you film in San Francisco, so I can help musically, for **one percent** of profits. The public is now curious enough to support ampu-actors in R-rated juicy love scenes, thanks to Heather McCartney! Please show this letter's first paragraph to the right people!

                                                                                                   Regards.

Letter from a 'film maker' – see page 134-136

interested in having a movie from me. One day he told me that he even had not a film camera. He was searching on Ebay a camera to make the film.

A same kind of request I received once from a Brazilian guy who wanted to make a movie with me in it. He told me that his script would be the way that I was a poor amputee, living in the streets. After taking me inside in his home, taking care for me, we would have sex. Everything on film, a gay adventure. For sure I wasn't interested in that.

I am not gay; I have nothing against gays, but for sure I wasn't again not interested in doing such a thing.

So many people are gay or bisexual. When we look now at the statistics of devotees, -I asked a good friend who knows also a lot of devotees and wannabes; Then this is the answer:

On 1000 people, the more recent evaluation is 2% = 20 devotees. Among them, in the totality of the European countries: 11% gays (this is official, E U) ; But in the great cities, like Brussels ,Paris or London : 20 to 22% are gays (men and females). The world record is in San Francisco : 34%. are gays.

There is a little way to know how many devotes or wannabes there are in the European community :

It is in the great rehab centers : the orthopedists are now attentive to this phenomenon, and it is this guys who think that 2% or perhaps more, are devotes, because they observe that, among their new amps, there are always a few guys who are accepting very well there amputation...

A big part of the devotees is indeed gay. I received a few times invitations by gays, worldwide.

I remember a story from a guy from California who invited me to go over there. He would pay everything. The travel, the stay... His wife would be out for a few weeks on a business travel, and he would offer me his nice vacation house he had near the coast, to stay in.

He had a private hospital, was specialized in dentistry. And after his hours of work. He would stay with me.

For sure a devotee who is bisexual. Also in this case I wasn't interested in the invitation. If I had accepted all invitations I ever received, I probably had seen the whole world.

There is a group of straight people, who become bisexual when it comes to disability.

A good friend from New York is haply married, lives on a boat, but when it comes to stumps, then he is interested in both genders.

If we would look deeper into the subject devotees, then there is a lot of information about them on the internet. For sure it belongs in the group of a pharaphilia.

Acrotomophilia is known as a pharaphilia of the stigmatic/eligibly type in which sexual erotic arousal and facilitation or attainment of orgasm are responsive to, and contingent on a partner who is an amputee.
Acrotomophilia was first recognized in clinics as rare phenomenon in the western world in the early 1990s.
Someone who has acrotomophilia (in the clinical sense) is sexually attracted to persons with an amputation, and may be un-attracted to people without amputations (although some can manage by fantasizing that their partner does have an amputation). That's being sexually aroused by an amputee, either in thoughts or actual deeds. Someone who has this fetish may even ask their partner to wrap one of their limbs in a bandage to fulfill their fantasy.
In its most extreme forms, a person my be unable to become aroused or achieve orgasm with anyone who does not have an amputation.
Therefore in apotemnophilia and acrotomophilia the desire for amputation may be a "counter phobic" antidote for males who fear castration.

Facts and Tips about Acrotomophilia

- Acrotomophilia disorder is an attraction to disability is a sexualized interest of people in the appearance, feeling and experience of disability.
- Acrotomophilia disorder lead to normal human sexuality into a type of sexual fetishism. Serologically, the pathological end of the attraction tends to be seen as a pharaphilia, though also as an aspect of identity disorder.
- There are most extreme forms, a person can be unable to become aroused or achieve orgasm with anyone who does not have an amputation.

At the time I had the site Ampulove online, I received a lot of stories from devotees who told me their story. Here are a few of them, without mentioning names.

-It was about the age of 8 or 9 (maybe earlier) that I began to have an obsessive curiosity for women/girls with disabilities. I was absolutely fascinated with them and attracted to them in the extreme. To narrow it even further, I especially like women/girls with limb-deficiencies. I fantasized at great lengths about how it might be to talk with them, know them, and be allowed to examine and fondle their disabled parts. (this has not abated through the years either)

My first arousing (and what I would call erotic) encounter happened about age 10 when the pretty young mother of a schoolmate came to our 4th grade class to show some home movies. The schoolmate had cautioned the class the day before that her mother had been born minus her left hand and wrist. I don't remember anything about the content of the home movies because I was completely enraptured with scrutinizing every movement she made as she deftly operated the projector with her right hand and left forearm stump. That was a sensual awakening for me, before puberty. My obsession with limb-deficiencies is almost primal! I can scarcely look back and never remember when I was not fascinated.

An accident I had at age 11 gave great impetus to this behavior. I fell from a piece of playground equipment from quite high up and landed on my right arm. The impact practically tore my right forearm off. The humerus had broken off above the elbow joint; the joint itself was shattered in 4 pieces and turned inside out. The wrist was broken in 2 places. My arm was attached only by the skin and flesh at the back of my elbow. When I staggered to my feet, I held my right arm straight out to examine it because I could not feel it, due to nerve trauma. Witnesses said my forearm swung vertically like a pendulum and from my vantage point all I could see was a bleeding stump with the white humerus bone protruding out. In horror, I thought I had really torn my arm off because I could not see it or feel it! I finally located and steadied my flopping right forearm with my left arm and what an eerie feeling because I had absolutely no sensation from my elbow down to my fingers! I spent 3 weeks in the hospital and went through several surgeries. At one point I came within just a couple of days of having my arm amputated because of a raging infection brought on by tanbark getting jammed into the compound fracture wound. I was a "guest" amputee for almost a year because I could neither bend my arm nor move it much because of nerve trauma. However, I did eventually recover almost full-use of my right arm. That was a frightening time for me and is the source of my extreme empathy for all people with disabilities ever since, especially amputees. "Wannabes" I simply cannot understand because teetering on the abyss of losing my right arm was sheer terror! God, I am lucky!

During my youth and young adulthood (about ages 17 to 25) I actively sought out disabled females to date and seek relationships with. I went steady with a girl blind from birth and another who had "Petite-Malls" epilepsy. However, I desired a young female amputee but it seemed that during that period, they had all been removed to another planet because I scarcely saw a single one that I could encounter, let alone date! At the age of 23, I met my future and present wife who was beautiful, sexy, had all her various parts, and was attracted to me, of all people. We courted, married, and the rest they say is "history." I would not be honest if I did not confess that one of the reasons I initially got married was to give my life a measure of "normality," something I knew in my heart it lacked. My wife turned out to be greater than the sum of her parts and we are still together after 27 years. Aside from my knowledge of God, she is the single most important person in my entire life.

For many years throughout my family life, I held my obsession/fascination with great secrecy. A clinical psychologist that I consulted briefly said I had a "shame-based" personality and a life-long low-grade depression. (He used a word for it that I could not even pronounce, let alone spell) Other than that, he was clueless and of little help. After many years of torment and anxiety about my abnormal preoccupation, I got onto the worldwide web for the first time about a year ago. Once on, I proceeded, as I always have, to do a search of "amputee" to see what was "out there." Much to my astonishment, I discovered that I was not alone in my fascination with limb-deficient females. There exists an entire vast subculture of like-minded people who share a similar interest. And I thought my unusual preoccupation was unique in all the world! My particular behavioral obsession, in clinical terms, is called "Amelotatism" or "Acrotomophilia," which in laymen's' terms means a fetish-like attraction/admiration/fascination with disabled women, especially those who are limb-deficient. Psychological research on this unusual behavior indicates it is acquired at a very young age, for many as young as four or five. My grandparents had two close friends, one with one arm and one with one leg and I was about four or five when I became acquainted with them They were both men but this could have been another impetus to this behavioral abnormality. Even though my orientation is still not universally socially acceptable, it is no more unusual for me than for a man to prefer blondes or heavy women or tall women or dark-complected women or any other very particular physical attribute. Aside from that, I feel much better about myself, but still have to deal morally with the accompanying feelings of erotic and sexual arousal when encountering limb-deficient females. Perhaps one of the reasons I have done volunteer/advocacy work for the disabled is out of remorse for exploiting their images in a manner that many of them would not condone. I do not collect pictures, videos, movies or any other images because I do not want to be discovered by my family. On the net, I look for images, appreciate them and move on. (It sure beats digging through the stacks of any health science library I happen to be near looking for female amputee images!) For me, there is nothing better than a live encounter with a disabled woman where I can initiate a two-way dialogue. I never leave one of these encounters without a word of encouragement or admiration because it is genuine. (By the way, my wife and two children know nothing of this and I aim to keep it that way. There is nothing to gain after all of these years for me to reveal any of this to my immediate family.)

Another story I received from a guy from 21 years old was about an amputee girl he had seen:

-Even I have been fascinated to disabled women since I was in junior high school in China. One of my intern teacher was a LAK amputee. She taught biology and she never use prosthesis. She said them were "clumsy". So she always let her empty sleeve swung in the class and use her only arm expertly. She told us she lost her arm due to some tissue disease when she was 17(she was 21 then). Something amazing about this cute lady is she was totally open to talk about her stump and even used it to whacked us.

She was a hot witch anyway. According to my friend, her stump made him "explore out" when them played baseball together. Now I don't know where she is, maybe married, even have kids. Many young teachers chased her was what I know. Her name was Li, which means "pretty" in Mandarin. The most exciting sight I've ever seen after was when I came to the city of Seattle of the United States. It was someday in July,1998. I went to the Woodland Park Zoo of Seattle. I walked down the path to the Codiac Bear exhibition hall and I was frozen---A wheelchair moved towards me. There was a slim women in the wheelchair, typical oriental face, but not young, maybe 33-36. Jade black hair and sexy blue sleeveless shirts with two China-like bare arms. Bright skin and pretty firm bosoms, maybe 33C. Her upper body was just perfect. But down to her waist, there was nothing but two empty pant legs hiked up by the wind. The end of her delicate stumps emerged a little, I could see it was a little bit crispy. But very fleshy and seductive. You can just bet how I wanted to touch them.

When she past me she nodded to me slightly and smiled. I was shocked to see she had a Chinese character tattoo on her shoulder, which says: "Tolerance." Jesus, she could be my compatriot! I tried to make a conversation with her but I didn't dare. Finally she disappeared in the tropical rain forest and let me stood there, try to imagine the feeling of having sex with such a charming legless lady.

An other story from a devotee who wrote me a few years ago and is together with a female amputee:

-I first realized that I was attracted to female amputees at the age of 8 when I saw a girl of 12 sitting on the ground outside a store. She was wearing a knee-length skirt and she had a peg-leg. I watched in fascination for a long as I could. I never saw her again - despite looking. There may have been other "sightings", but the next one which is indelible in my memory was at the age of 13. I used to catch the bus to school. On Wednesday we played sport and I could catch an earlier bus home. At 3.20 pm the bus pulled up at a particular stop and a woman got on. She was in her late 20s. She sat in the front of the bus. She reached into her purse with her left hand and pulled out the appropriate fare. The coin was pressed between her thumb and middle finger. The index finger was missing just below the outer joint. She was a 4.1.8 (4th finger, second joint, 80% left). This woman fascinated me. I caught the same bus every Wednesday for as long as I could until one Wednesday she didn't show up. I never saw her again. I have witnessed many sightings of women with a missing finger (or fingers) since that time.

It is neatly amputated fingers that most interest me in women. I like, say, a missing leg or a missing arm, but the fingers, one or more, have it. I particularly like an index finger, preferably with some of the finger left.

Like others I felt as if I was the only one in the world who had this fascination which definitely has taken its toll on my emotions since, certainly, 13 years old. I had felt this way until I searched the internet and discovered others for the first time some years ago. I certainly feel relieved that others share my feelings. To me it is perfectly natural to see a missing part or parts comprising one's preference. The woman must have other qualities, but the missing finger acts as one of her unique characteristics.

I guess my story has a happy ending. I am now married very happily to a finger amputee. She enjoys my interest although I am sure, at times, that she still finds it somewhat mystifying.

And let me offer a last story of a devotee:

-I have been attracted to the disabled for as long as I can remember. I would embarrass my mother by trying to peek up the cutoff pant leg of a SAK. I remember seeing a little girl around my age, maybe older at a Sears department store, who wore a prosthetic hook. I asked my father how I could get a hook arm, and he replied to me, "Only if you're very unfortunate." I had no shame as a kid, it would seem to me as I looked back at my lifelong fascination. I asked a day-camp counselor how someone ended up with no legs and in a wheelchair. She mentioned something to me about "gain-green". I grew up during the post-Vietnam era and remember seeing a guy in some public service announcement with a disabled vet with no legs rolling through his workplace. I also remember seeing a show on TV about a woman with major limb deficiencies, using the remainder of her fingerless hands to type. She too was in a wheelchair and had no legs.

People who were in wheelchairs and otherwise whole also fascinated me from a young age, especially if they were wearing socks and no shoes. I've seen people with muscular dystrophy, their feet so amorphous that shoes were a foregone conclusion. Jumping to my college years, I attended the University of Illinois at Urbana-Champaign.

Now, if ever there was a college for the devotee, this may have been it. Now, there wasn't a preponderance of students in wheelchairs, but definitely more than most colleges. It is probably the most wheelchair-accessible campus in the country, situated perfectly in the flatlands of East Central Illinois--hardly any hills to speak of. While at U of I, I got to see my share of fairly attractive young paraplegic women, sometimes not wearing shoes. There was this one gal in a chair who would roll around, one leg crossed over the other leg, and for some reason only known to this gal, the foot of her crossed leg was always shoeless, and she'd be wearing a sock. Another woman who I would sight from time to time was a SAE who wore a prosthetic hook most of the time. I do recall one time sighting her without her hook while I was riding in a car. It does need to be noted that I was very much in the closet about my attraction to disabled women while in college. If I had the chance to meet a disabled woman there (which I didn't... they seemed not to travel in the same circles as me, being a loose-cannon campus radical and an embarrassment to most of the campus radicals at that...). Out of college, and being true to my underachieving history, I got a job as a bike messenger in downtown Chicago. I did get to see my share of walking-disabled/orthopedically-challenged individuals: women wearing those open-toed booties, beggars missing legs, paraplegic professionals. Just to make it perfectly clear, I didn't take the job of a bike messenger just to gawk at cripples downtown. You see, I was determined not to take what I thought of as a corporate sellout job that would make me cut my (then shoulder-length) hair. I wanted to wage class war as a member of the working class--or at least in a working-class-type job and organize the messengers so we could rise up and make revolution... But I could scarcely organize my own life, much less my workplace.

The stories I describe are coming from man. I had a very beautiful story from a female devotee and wannabe a few months ago. She's from the Netherlands. During a few years we are in contact with each other. She want to become legless, but she is more a devotee then a wannabe.

The story she wrote down, was about her becoming legless, but with me involved in it. How we both would go through live as an amputee, each in their own wheelchair. A very nice story, but after a computer crash I had... I lost the possibility to publish it here.

Near the Dutch girl, there was a French woman. Sandrine is her name. I received an email from her to meet me:

> Hi Alex
>
> I thought about what you told me. I think I have feelings for you.
>
> I might come to the meeting that you organize in august.
>
> I wanted to ask you one thing.
>
> Do you think that I could get may amputations during this meeting in Miami ?

> It is just small amputations to start , in my case.
>
> Sandrine with love to you Alex

An email from April 7, 2008. That time I was planning indeed a meeting between amputees and devotees, here in Miami, FL. After more conversations with Sandrine, it was sure she was female, devotee and wannabe.
Another female devotee I am in contact with is Renee. She is 18 years old and attracted to male amputees. Here is one of the most recent emails she send to me, after inviting me to China:

> Dear Alex,
>
> Happy for your reply.
>
> If you will go travel with me to Yangshou, please forgive me that I can't let my parents know you, cause they must think I'm crazy and wouldn't let me out.
>
> Sitting with you in Yangshuo is so romantic
>
> Are you ex-wife a devotee as me?
>
> And I haven't think about how to talk to my parents, my mother never consider if I would have a relationship with amputee

guy, and when she heard of someone married a foreign boy, she always said "You should marry a Chinese boy". But if we fall in love, I think I will persuade them, but maybe it takes many years.

You have very beautiful stumps.

Hope your reply.

Yours,

Renee.

A few months ago, I heard some strange stories about Chinese devotee man, that would pay woman to become an amputee. But more people started to talk to me about it. It sounds like some man in China can't find the perfect amputee partner. They can find the perfect woman, who isn't an amputee, and offer this woman a lot of money to become an amputee.
Unfortunately there isn't a lot of information from it on the internet, except in closed wannabe groups.

On devotees I did a few investigations through the internet.
The questions I asked:

01. your gender
02. your age
03. your sexual preference
04. How old where you, when you find out that you are attracted to amputees
05. to what gender of amputees you are attracted
06. What amputation do you adore the most
07. Describe why you find amputees attractive
08. Did you ever met an amputee
09. Where you ever in a relationship with an amputee
10. When you would meet an amputee, would you tell him/her about your devotee feelings
11. Are you also an amputee
12. Are you also a wannabe
13. Did you ever talked to someone about your admirer feelings (family, friends, doctor...)
14. In what kind of relationship are you now
15. If you are/where in a relationship with a non-amputee, and you would meet an amputee, would you stop your relationship to get further with the amputee

16. If you would meet an amputee, and she or he tells you that she or he was an amputee by choice (ex-wannabe), would you still went on with meeting her/him ?
17. Tomorrow you read in the local newspaper that a male / female amputee prostitute is looking for contact, and asks 50$ for one hour, would you phone her / him for a meeting
18. What would you do if you had the possibility to meet an amputee, but she have not the amputation you prefer
19. Is the shape of a stump is important for you
20. Is the size of a stump important for you
21. Did you ever met an amputee trough an organization like the Russian Frantana, and Amputy
22. Would you start a relationship with a ugly amputee, but someone who have the amputation you prefer
23. Are you only interested in Caucasian amputees, or also in African amputees

24. How important is it for you to meet or be with an amputee for real, on a scale of 10
25. Do you prefer a prosthetic wearing amputee, or someone who don't use prosthesis

All results arrived anonymous. Here are a few of the most interesting results, I received from female devotees. The answer, located near the number of the question.

answer01: female
answer02: 50
answer03: straight
answer04: 16
answer05: female amputees
answer06: right above elbow
answer07: look like interesting and pretty
answer08: yes
answer09: yes
answer10: yes
answer11: yes
answer12: no
answer13: no
answer14: married
answer15: no
answer16: no
answer17: no
answer18: I would meet him/her

answer19: no
answer20: no
answer21: no
answer22: no
answer23: only Caucasian
answer24: 3 on 10
answer25: This is not so important for me

---

answer01: female
answer02: 42
answer03: straight
answer04: 24
answer05: male amputees
answer06: double hip disarticulation
answer07: I don't know what, but for me a double above the knee amputee with a well built upper body is the most erotic.
answer08: no
answer09: no
answer10: I don't know
answer11: no
answer12: no
answer13: no
answer14: married
answer15: no
answer16: yes
answer17: no
answer18: I would meet him/her
answer19: no
answer20: no

answer21: no
answer22: no
answer23: black or white, I don't care
answer24: 6 on 10
answer25: Sometimes prosthetics, but most of the time without

---

answer01: female
answer02: 43
answer03: straight
answer04: 10
answer05: male & female amputees
answer06: double hip disarticulation
answer07: I don't know
answer08: yes
answer09: yes, and I still am
answer10: yes
answer11: no
answer12: no
answer13: no
answer14: married
answer15: yes
answer16: yes
answer17: no
answer18: I would meet him/her
answer19: yes
answer20: no
answer21: no
answer22: no
answer23: black or white, I don't care

answer24: 1 on 10
answer25: This is not so important for me

---

answer01: female
answer02: 37
answer03: straight
answer04: 8
answer05: male amputees
answer06: quad amputees (4 limbs)
answer07: I find it a nice idea to take care for a man who can't do anything more
answer08: yes, I met already an amputee
answer09: no, I never was in a relationship with an amputee
answer10: yes, I will tell him/her that I am admired to amputees
answer11: no, I am not an amputee
answer12: no, I am not a wannabe
answer13: no, I never talked to someone about my admirer feelings
answer14: separated / divorced
answer15: yes, I would leave my partner, and start a new live with the amputee
answer16: yes, I would go on with meeting that ex-wannabe
answer17: no, I would not contact a amputee prostitute
answer18: I would meet him/her, even if he/she don't have the amputation I prefer

answer19: no, the shape of a stump is not important for me
answer20: yes, the size of a stump is important for me
answer21: no, I never met an amputee trough an organization
answer22: no, I would not meet, an amputee need to be beautiful
answer23: black or white amputees, I don't care
answer24: 10 on 10
answer25: I prefer an amputee without wearing prosthetics

---

answer01: female
answer02: 35
answer03: straight
answer04: 12
answer05: male amputees
answer06: quad amputees (4 limbs)
answer07: I don't know
answer08: yes, I met already an amputee
answer09: no, I never was in a relationship with an amputee
answer10: I don't know or I will tell that I am an admirer
answer11: yes, I am also an amputee
answer12: I am an ex-wannabe who is an amputee
answer13: no, I never talked to someone about my

admirer feelings
answer14: single
answer15: yes, I would leave my partner, and start a new live with the amputee
answer16: yes, I would go on with meeting that ex-wannabe
answer17: no, I would not contact a amputee prostitute
answer18: I would meet him/her, even if he/she don't have the amputation I prefer
answer19: no, the shape of a stump is not important for me
answer20: yes, the size of a stump is important for me
answer21: no but I am planning to meet an amputee trough an organization
answer22: no, I would not meet, an amputee need to be beautiful
answer23: I am only interested in Caucasian amputees
answer24: 8 on 10
answer25: I prefer an amputee with prosthetics

---

answer01: female
answer02: 24
answer03: straight
answer04: 12
answer05: male & female amputees
answer06: left above knee

answer07: I like stumps and the helplessness
answer08: yes, I met already an amputee
answer09: yes, I was already in a relationship with an amputee
answer10: I don't know or I will tell that I am an admirer
answer11: no, I am not an amputee
answer12: yes, I am also a wannabe
answer13: yes, I talked already to someone about my devotee feelings
answer14: separated / divorced
answer15: yes, I would leave my partner, and start a new live with the amputee
answer16: yes, I would go on with meeting that ex-wannabe
answer17: yes, I would contact that amputee prostitute
answer18: I would meet him/her, even if he/she don't have the amputation I prefer
answer19: yes, the shape of a stump is very important for me
answer20: no, the size of a stump is not important for me
answer21: no, I never met an amputee trough an organization
answer22: no, I would not meet, an amputee need to be beautiful
answer23: black or white amputees, I don't care
answer24: 10 on 10
answer25: I prefer an amputee without wearing prosthetics

answer01: female
answer02: 18
answer03: straight
answer04: 16
answer05: male amputees
answer06: right foot amputation
answer07: I don't no
answer08: no, I never met an amputee
answer09: yes, and I still am in a relationship with an amputee
answer10: I don't know or I will tell that I am an admirer
answer11: no, I am not an amputee
answer12: I am an ex-wannabe who is an amputee
answer13: yes, I talked already to someone about my devotee feelings
answer14: married
answer15: no, I would not leave my partner for the amputee
answer16: yes, I would go on with meeting that ex-wannabe
answer17: no, I would not contact a amputee prostitute
answer18: I would not meet an amputee who don't have the amputation(s) I like
answer19: yes, the shape of a stump is very important for me
answer20: no, the size of a stump is not important for me
answer21: yes, I met once an amputee trough an

organization
answer22: no, I would not meet, an amputee need to be beautiful
answer23: I am only interested in Caucasian amputees
answer24: 8 on 10
answer25: Sometimes prosthetics, but most of the time without

---

answer01: female
answer02: 21
answer03: straight
answer04: 12
answer05: female amputees
answer06: triple amputees (3 limbs)
answer07: Whenever I see an amputee I get sexually aroused, and my head starts to fantasies about in what way I would have sex with her(the things we would do to each other). I don't know why I am a devotee.
the first time I saw an amputee I just stared and I was completely intrigued about the amputation, how it is to live with no legs or hands or other combinations and started to fantasies about it.
I think I'm going to be a Wannabe in the future.
answer08: yes, I met already an amputee
answer09: no, I never was in a relationship with an amputee
answer10: yes, I will tell him/her that I am

admired to amputees
answer11: no, I am not an amputee
answer12: yes, I am also a wannabe
answer13: no, I never talked to someone about my admirer feelings
answer14: single
answer15: no, I would not leave my partner for the amputee
answer16: yes, I would go on with meeting that ex-wannabe
answer17: yes, I would contact that amputee prostitute
answer18: I would meet him/her, even if he/she don't have the amputation I prefer
answer19: no, the shape of a stump is not important for me
answer20: no, the size of a stump is not important for me
answer21: no, I never met an amputee trough an organization
answer22: yes, I would meet her/him, I don't care how she/he looks like
answer23: I am only interested in Caucasian amputees
answer24: 6 on 10
answer25: Sometimes prosthetics, but most of the time without

---

answer01: female
answer02: 30
answer03: homosexual
answer04: 15
answer05: female amputees
answer06: double hip disarticulation
answer07: I love the idea to take care of an amputee
answer08: no, I never met an amputee
answer09: no, I never was in a relationship with an amputee
answer10: no, I will not tell that I am a devotee
answer11: no, I am not an amputee
answer12: no, I am not a wannabe
answer13: no, I never talked to someone about my admirer feelings
answer14: single
answer15: yes, I would leave my partner, and start a new live with the amputee
answer16: yes, I would go on with meeting that ex-wannabe
answer17: no, I would not contact a amputee prostitute
answer18: I would meet him/her, even if he/she don't have the amputation I prefer
answer19: yes, the shape of a stump is very important for me
answer20: yes, the size of a stump is important for me
answer21: no, I never met an amputee trough an organization

answer22: yes, I would meet her/him, I don't care how she/he looks like
answer23: black or white amputees, I don't care
answer24: 7 on 10
answer25: I prefer an amputee with prosthetics

---

answer01: female
answer02: 19
answer03: straight
answer04: 14
answer05: male amputees
answer06: left hip disarticulation
answer07: they are awesome
answer08: no, I never met an amputee
answer09: no, I never was in a relationship with an amputee
answer10: no, I will not tell that i am a devotee
answer11: no, I am not an amputee
answer12: no, I am not a wannabe
answer13: no, I never talked to someone about my admirer feelings
answer14: single
answer15: no, I would not leave my partner for the amputee
answer16: no, I would not go on with meeting a ex-wannabe
answer17: no, I would not contact a amputee prostitute

answer18: I would meet him/her, even if he/she don't have the amputation I prefer
answer19: no, the shape of a stump is not important for me
answer20: no, the size of a stump is not important for me
answer21: no, I never met an amputee trough an organization
answer22: no, I would not meet, an amputee need to be beautiful
answer23: I am only interested in Caucasian amputees
answer24: 1 on 10
answer25: This is not so important for me

---

answer01: female
answer02: 16
answer03: straight
answer04: 10
answer05: male & female amputees
answer06: double below elbow
answer07: amputees are sexy!!!!
answer08: no, I never met an amputee
answer09: no, I never was in a relationship with an amputee
answer10: yes, I will tell him/her that I am admired to amputees
answer11: no, I am not an amputee
answer12: yes, I am also a wannabe

answer13: no, I never talked to someone about my admirer feelings
answer14: living together
answer15: no, I would not leave my partner for the amputee
answer16: yes, I would go on with meeting that ex-wannabe
answer17: no, I would not contact a amputee prostitute
answer18: I would meet him/her, even if he/she don't have the amputation I prefer
answer19: no, the shape of a stump is not important for me
answer20: no, the size of a stump is not important for me
answer21: no, I never met an amputee trough an organization
answer22: no, I would not meet, an amputee need to be beautiful
answer23: black or white amputees, I don't care
answer24: 7 on 10
answer25: I prefer an amputee without wearing prosthetics

---

answer01: female
answer02: 33
answer03: bisexual
answer04: 30
answer05: male & female amputees

answer06: quad amputees (4 limbs)
answer07: Amputees interest me. I want to know how they live their lives and what they're interests are. I am not just interested in them because I think they are sexually attractive. I want to find out what's in their minds. As I am most attracted to quad amputees, I want to take care of their every need.
answer08: yes, I met already an amputee
answer09: no, I never was in a relationship with an amputee
answer10: no, I will not tell that I am a devotee
answer11: no, I am not an amputee
answer12: yes, I am also a wannabe
answer13: yes, I talked already to someone about my devotee feelings
answer14: married
answer15: no, I would not leave my partner for the amputee
answer16: yes, I would go on with meeting that ex-wannabe
answer17: no, I would not contact a amputee prostitute
answer18: I would meet him/her, even if he/she don't have the amputation I prefer
answer19: no, the shape of a stump is not important for me
answer20: no, the size of a stump is not important for me
answer21: no, I never met an amputee trough an organization

answer22: yes, I would meet her/him, I don't care how she/he looks like
answer23: black or white amputees, I don't care
answer24: 7 on 10
answer25: I prefer an amputee without wearing prosthetics

---

answer01: female
answer02: 53
answer03: straight
answer04: 25
answer05: male amputees
answer06: single above knee
answer07: I am attracted sexually, I get turned on by the sexual
possibilities. Also attracted by their courage to live with this condition.
answer08: no, I never met an amputee
answer09: no, I never was in a relationship with an amputee
answer10: yes, I will tell him/her that i am admired to amputees
answer11: no, I am not an amputee
answer12: no, I am not a wannabe
answer13: no, I never talked to someone about my admirer feelings
answer14: separated / divorced
answer15: no, I would not leave my partner for the amputee
answer16: no, I would not go on with meeting a ex-wannabe
answer17: no, I would not contact a amputee prostitute
answer18: I would meet him/her, even if he/she don't have the

amputation I prefer
answer19: yes, the shape of a stump is very important for me
answer20: yes, the size of a stump is important for me
answer21: no, I never met an amputee trough an organization
answer22: yes, I would meet her/him, I don't care how she/he looks like
answer23: black or white amputees, I don't care
answer24: 7 on 10
answer25: Sometimes prosthetics, but most of the time without

---

answer01: female
answer02: 28
answer03: straight
answer04: 15
answer05: male amputees
answer06: right hip disarticulation
answer07: I saw amputee in TV.I felt something in my mind but I don't know what does it mean. I know everything on Internet
answer08: yes, I met already an amputee
answer09: no, I never was in a relationship with an amputee
answer10: no, I will not tell that i am a devotee
answer11: no, I am not an amputee
answer12: no, I am not a wannabe
answer13: no, I never talked to someone about my admirer feelings
answer14: single
answer15: no, I would not leave my partner for the amputee
answer16: yes, I would go on with meeting that ex-wannabe
answer17: yes, I would contact that amputee prostitute

answer18: I would meet him/her, even if he/she don't have the
amputation I prefer
answer19: yes, the shape of a stump is very important for me
answer20: yes, the size of a stump is important for me
answer21: no, I never met an amputee trough an organization
answer22: yes, I would meet her/him, I don't care how she/he looks like
answer23: I am only interested in Caucasian amputees
answer24: 7 on 10
answer25: Sometimes prosthetics, but most of the time without

In a few of the inquiries it is clear that some devotees are interested in the fact to take care for someone that is helpless. Not only female devotees, but also males.
A lot of female devotees work also in the medical world and are nurse. When I had my right below the knee amputation, the nurse at night, she was a devotee too.

She only worked at night, was very pretty, and I guess around the age of 25. A lot of nights she came to sit with me on the bed, brought me even an ashtray, allowed me even to smoke in the hospital room, and even brought me some beers. The way to kissing each other for a lot of minutes was close...

After the hospital period, she wanted to date me. She didn't lived that far from my place, but it never came to a meeting outside the hospital.

What the most devotees prefer isn't very clear. Some prefer short stumps, but there are admirers that don't like stumps. A friend of me like hip amputees, without stumps.
Coming to the stump, mostly the stump is important. Someone even told me once: "The stump is for me a reincarnation of a breast". With other words, he was saying that a stump was for him like a breast of a woman.

Again, we may not forget that devotees are not only amputation oriented; Only the biggest part is.
And some devotees are a wannabe, or even become a wannabe.
As an example: Someone is devotee, fall in love with an amputee. After a while the devotee, -who as the time to observe the living style of the amputee partner, can get so interested in it, that he or she want to be in the same situation. A lot of times, seeing that an amputee have much more attention can bring this idea up, to become an amputee also.
In this way, it can be that a wannabe discover his first feelings when he or she

is together with an amputee, first as a devotee.

I could go on with writing this book, in a way that it would become a (medical) encyclopedia, or even a story book, but basically I am only interested to tell only what I found out, what I know.

About the amputees I met, there are different feelings around devotees and wannabes. Some amputees can accept that there are wannabes, but the most of them don't understand why they want to have a limb removed. About devotees; First devotees were not so much appreciated by amputees. The last years, this changed a lot. Those days, the most amputees are not longer upset about the fact there are admirers. I know a lot of amputees who are in a happy relationship with an amputee and that know that their partner is an acrotomophile.

The next and last part of the book will go pure about me and my opinion around the whole subject.

* * *

My opinion and more around me

Already a lot of information was given in this book about me, but there is always more to tell.
When I had this morning a friend on Skype, who is a wannabe, he told me: 'Make the book with also a few pictures of you in'.
I could had write so much more stories, examples, medical information, but my idea wasn't to make a kind of never ending books, -or medical oriented books.
More about my opinion, idea around the devotee and wannabe world.
Almost fifteen years I am busy with the phenomenon's devotee and wannabe, and everyday there is more to discover.
A lot of people asked me already or I was already together with an amputee.
Yes, I was.
When I had years ago the website Ampulove, I met amputees, interesting in modeling. So also an English woman who is legless and armless.
While we were making pictures and movies, -I guess we felt in love. We shared together an hotel room, her parents invited me, we went out together.

When afterwards, being back in Belgium, she contacted me with the request to live together with her in London; I said 'No'.
That time, I wasn't interested in living in another country. Never thought I would even live in another country like I am doing now.
Since a good four years ago I live here close to Miami in Florida, USA.

A lot of times when I go out, I can observe very well the devotees, for sure the woman. Once I went out with a friend, there were two very beautiful woman walking. That was during a weekend in Bayside, Miami. For sure they couldn't keep their eyes away from me. An other time, in the city of Fort Lauderdale downtown (Las Olas); There was a male amputee, no arms, and with him together, a very beautiful girlfriend.
'Is that a devotee?', is what I was thinking.
For myself, I consider the biggest group of woman I was with, since I am an amputee, - a devotee.
Maybe not a 100 % devotee, but for sure a woman that have no any problems with being together with an amputee. Basically a woman that like it to be together with

someone who have stumps, a legless guy who don't walk anymore.
Or is it maybe because I am so open minded and make jokes about it?
'Why should I cry with the fact I am an amputee?', is mostly my reaction anyway.
Once I was in an airport in Brazil. There was someone that cleaned shoes for money. I went to him, asked him how much to pay for my shoes.
He was looking at me, and really didn't know how to react on what I just told him.
Another time, also on an airport. I went for a ticket.
When the person of the agency told me the price to pay, I reacted: 'I pay only 50%, I will travel only 50%'; With the fact I am legless and have one arm.
Once, here in Florida. I went together with my best friend to Wal-Mart. The cashier was complaining about pain in her knees, from standing a lot during the whole days.
My reaction was: 'I have that also a lot'.
She looked at me, started to laugh, and left for a minute.
Two years ago, after waking up, I asked a girlfriend:
'Sweetie, can you please bring my shoes too?', while she was taking her shoes.
I think the poor girl wasn't yet good awake.

'Where are your shoes?', she asked me, knowing that I have no legs.
'Oh God... You idiot!', she answered a second later, finally realizing that my legs are gone and that I don't wear shoes.
When I have someone in front of me that feels really sorry for me that I am an amputee, then I mostly answer:
'Well don't be, my live is much cheaper now'
Their question?
'Well, I don't need to buy shoes; You have any idea how expensive shoes are those days?', is then the answer.
Mostly they start to laugh afterwards and realize that they can be themselves without being shy.
I am for sure not a shy person.
A few years ago, in the South of France, -I was on the beach; A man came to me, asked me very friendly or his daughter could ask me a few questions around my amputations.
He told me that she had observed me the whole time on the beach, and she wanted to know how I became an amputee.
A few later the daughter, -I guess she was around ten years old, was standing in front of me.

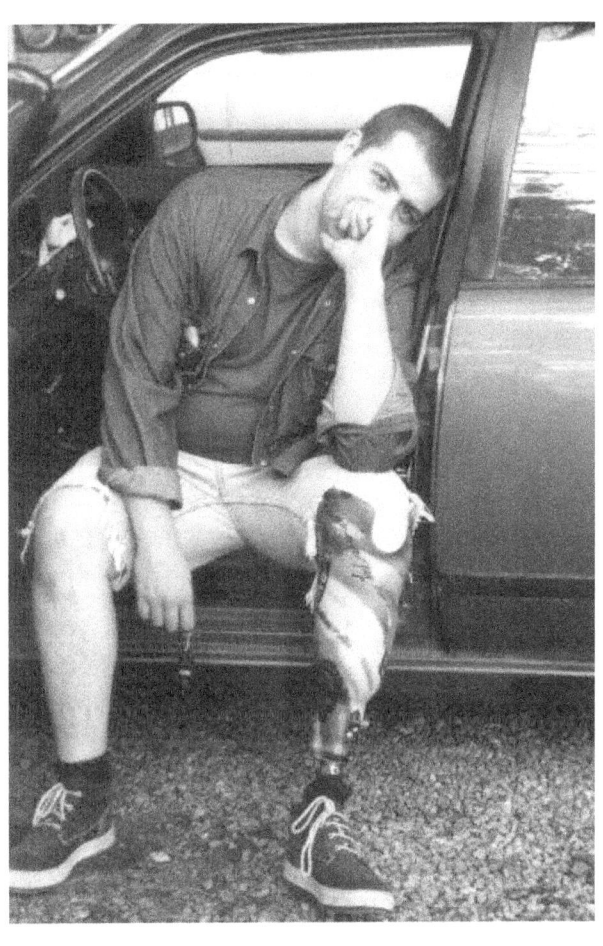

Alex as a left below the knee amputee years ago

After the question, I first didn't know what to answer her. Soon I came up with:
'Well, I was in the movie –the invisible man. The producer found a way to make me invisible, but not yet a way to bring me back and visible again'.
For me it was the best answer I could had give to someone from her age. Would had been to difficult to talk about accidents.
When people that don't know me, or don't know I am an amputee and ask me where I live?
'Well, one part of me is in Europe, the other part is in the USA'.
Off course they don't get it from the first moment. Then I answer them that I became an amputee in Europe, that my legs and my arm is still there, but that I moved to the USA.
A last funny one?
Yesterday evening. I went out with my girlfriend and showed the layout of a book cover I painted to one of her friends.
I received a question: Do you all paint that with the hand?
Off course the intention was to answer between, with the hand or with the computer.
My answer on this was: '... No, with my feet'.

And off course a few people started to laugh. Maybe it can be a little bit sarcastic, but so many people like sarcastic jokes.

So, I was once in a meeting with a female above the knee amputee, we went out together, went on restaurant together, and for sure the interest started to grown between us. When she heard about all the things I found out. That there are wannabes, that there are devotees, she wasn't interested anymore and we lost contact. Probably she started to freak out and thought maybe that I was interested in her for her stump, or maybe that I was nuts.

On all my worldwide travels, I have observed a lot of wheelchair accessible situations. Coming to Europe,... sorry. There they still think that a wheelchair user belongs to the group of people that need to stay inside the house. Same in South America. To find a wheelchair accessible restroom? Better do your needs in the streets then going further searching...
The best organization I ever have seen? Here in the USA. Everything need to be wheelchair accessible.

When I came to the USA about four year ago, I discovered that really everything here is wheelchair accessible. Every store, big or small; restaurants, bars, banks, ... eve-ry-thing is totally wheelchair accessible. And for sure there are more ramps here then streets. Every street have ramps.
Also people look different into the direction of people with a disability. Walking over the streets; -In my case rolling over the streets; No one looks at you, while I could feel in Europe and other countries I've seen the eyes watching in my back. I mean with this: People that don't steer at you at the moment you see them, but people that watch you when they think that you can't see they are looking.

In South America I had the weirdest reactions. People that think that a person with a disability isn't able to make love. No? –Once there was a woman who directly asked me the question: 'Alex, can you still have sex?'.
A little bit frustrated I answered:
'Easier then someone with two legs, at least I don't need to open my legs anymore!'.

Later: Alex as a left above the knee amputee

But you get off course funny reactions. Once on the street here, on my direction to the mall.
A man greeted me very friendly and said: 'Welcome home sir', probably thinking that I came back from a war.

The funniest ever was, going out with a friend to a bar. We were sitting outside and we started to talk about the subjects amputees, devotees, stumps and wannabes. Thanks to the drinks we had, we didn't realized how many people were watching us.

The saddest ever was when I became a right above the knee amputee. That time a friend from the Netherlands came to visit me who was a wannabe. He needed to see me with my new amputation, while that was what he absolutely wanted to have. Happy I was when I discovered a few months ago that he reached his goal, and became a right above the knee amputee.
So, my opinion about wannabes is very simple. I think that everyone on this planet have the right to be whoever they want to be. Maybe it is strange that someone is an apotemnophile or an acrotomophile; -Maybe strange because

you don't ear about it everyday. But if really someone is so unhappy, because he or she want to become an amputee, then that need to be possible. It is sad enough that some wannabes do such a bad things to themselves some even lost their live from doing dangerous things. Then, for sure my opinion is, it would had been better if she or he had became an amputee.

A wannabe told me once: 'I had first one leg off, no I am legless, so far I went'.

I answered him: 'So?', not knowing what to answer.

He reacted: Well, if I wasn't satisfied with my first amputation, would I had done afterwards another amputation?'.

He had indeed a point.

If he wasn't happy with the results of his first amputation, for sure he never had gone through a second leg amputation.

When I come to the point of devotees. For sure I like them. When a devotee, in my case mostly female devotees start to contact me, I always try to find out why they find me and amputations are so attractive for them. Basically I know already the most answers, but it is always nice to ear more. Everyone has different answers.

While I know that so many articles from me came out on this planet, I can still say that I don't mind. Some articles are really good, some are so crazy as it can be. But I really don't mind about it. There are so many sensation writers. Someone who understand what is sensation will use their brains when they find such an articles.

This week I did a lookup at Google. On Google only there are about 3500 links from me, or better about me.

When I went till page 10 on Google, I had already enough of it. Page 11, 123, 13, 14 and further probably had the same kind of links; Alex the amputee, Alex the wannabe, Alex and his wife in Florida, movies on YouTube, pictures and all what could be interesting for people into freaky articles.

For that reason, I mostly refuse any interview. Since I found out that the most interviews will be totally changed afterwards. I mean; I can say a 'yes', and afterwards they can bring it as a 'no'.

Only for serious interviews I still can be open. But how you know in advance that an interview will be serious or not?

**Now: Alex as a legless
and one armed amputee**

A few people asked me, why I took years ago Ampulove offline, why I don't bring back that site. If I ever would come back on the internet with a kind of Ampulove, then it would be only with information; - but basically all what I wrote down in this book is what I know. But helping further devotees and wannabes, to listen to them? To be there for them as a friend who is an amputee and very open minded for those subjects; And maybe, I am indeed so open minded and interested in and about wannabes that people may call me a wannabe too. After becoming an amputee, I never minded about. Legs or not, live goes on anyway.

And the most friends I have are anyway amputees, devotees and wannabes.

I almost can say that it is my family.

Alex Mensaert

www.ingramcontent.com/pod-product-compliance
Lightning Source LLC
Chambersburg PA
CBHW060846170526
45158CB00001B/253